The Price of Nuclear Power

Nature, Society, and Culture

Scott Frickel, Series Editor

A sophisticated and wide-ranging sociological literature analyzing nature-society-culture interactions has blossomed in recent decades. This book series provides a platform for showcasing the best of that scholarship: carefully crafted empirical studies of socio-environmental change and the effects such change has on ecosystems, social institutions, historical processes, and cultural practices.

The series aims for topical and theoretical breadth. Anchored in sociological analyses of the environment, Nature, Society, and Culture is home to studies employing a range of disciplinary and interdisciplinary perspectives and investigating the pressing socio-environmental questions of our time—from environmental inequality and risk, to the science and politics of climate change and serial disaster, to the environmental causes and consequences of urbanization and war-making, and beyond.

The Price of Nuclear Power

Uranium Communities and Environmental Justice

STEPHANIE MALIN

Rutgers University Press

New Brunswick, New Jersey, and London

Library of Congress Cataloging-in-Publication Data
Malin, Stephanie A., 1981–
The price of nuclear power : uranium communities and environmental justice / Stephanie A. Malin.
pages cm. — (Nature, society, and culture)
Includes bibliographical references and index.
ISBN 978-0-8135-6979-6 (hardcover : alk. paper) — ISBN 978-0-8135-6978-9 (pbk. : alk. paper) — ISBN 978-0-8135-6980-2 (e-book (web pdf))
1. Nuclear industry—United States. 2. Nuclear power plants—Environmental aspects—United States. 3. Nuclear industry—Information services—Government policy—United States. I. Title.
HD9698.U52M3195 2015
333.792'40973—dc23
2014030639

A British Cataloging-in-Publication record for this book is available from the British Library.

Visit our website: http://rutgerspress.rutgers.edu

Manufactured in the United States of America

This book is dedicated to uranium community members living across the desert southwest. Thank you for sharing your stories and allowing me to tell them here.

They're not going to save our community with an organic farm. Let us have our [Piñon Ridge Uranium Mill], and maybe we can afford your organics.
—Cindy, Nucla, Colorado

Why are [the uranium communities] Nucla and Naturita so distraught and poor and so devastated? The only reason I can see is because of uranium, and the idea that uranium is going to come in and make everything fine seems to me extremely questionable.
—Clint, Paradox Valley, Colorado

It may seem surprising, but we would support uranium industries if they expanded again in the area. All along, we've been told, "It's safe, it's safe." And all we can do is go on their word that it's safe. Who knows? Nobody knows what is safe. It's the risk we take.
—Fritz, Monticello, Utah

Contents

Illustrations

Maps

Photographs

Acknowledgments

My work in this book has depended on the generosity of the many organizations, mentors, colleagues, family, friends, and generally positive people in my life. I feel enormous gratitude for all of you.

I thank Rutgers University Press for having faith in my work and for publishing this vital but invisible story. Thanks also to Peter Mickulas for his expertise and support and to Scott Frickel, editor of the press's Nature, Society, and Culture series, for including this book. Thank you to Dawn Potter for her careful and considerate copy editing and gentle reminders to let the story tell itself.

I am grateful to the Rural Sociological Society, which offered me a doctoral dissertation award to support my fieldwork. With that assistance, along with a dissertation fellowship from Utah State University's Graduate School, I conducted extensive mixed-method, ethnographic data collection that enabled me to undertake high-quality historical-comparative research. Without their institutional support, my book would not exist.

I thank Utah State University's Department of Sociology, Social Work, and Anthropology, where I learned how to conduct conscientious, community-oriented, ethical sociology. Peg Petrzelka, I owe you a lifetime of gratitude for teaching me how to be a balanced and strong female academic. You have shown me how to state my case bravely and to channel humanitarian ideals into grounded sociological observations—while also striving not to overdo. Christy Glass, thank you for showing me how to be a passionate advocate for rigorous sociology and for your inspirational teaching. John Allen, Rick Krannich, Doug Jackson-Smith, Jon Morris, Sandy Marquardt-Pyatt, Susan Mannon, and Michael Toney: thank you all for teaching me how to work with and for the people and communities I have written about here. Rebecca Smith, Brian Jennings, Anita Harker Armstrong, Joyce Mumah, Stephen van Geem, Matt Cottrell, and my other graduate school friends: thank you for trips to the campus coffee shop, lively discussions, lots of laughter, and years of warm gatherings during snowy Logan winters. Becca and Brian, thank you for continuing to treat Matt and me like family.

I am grateful for generous support from Brown University's Center for Environmental Studies, its Cogut Center for Humanities, its Superfund Research Program, and its Department of Pathology and Laboratory Medicine. The Cogut Center's Mellon Postdoctoral Fellowship let me work with fabulous scholars, gave me time to write this book, and allowed me to teach engaged and inquisitive undergraduates. Phil Brown, thank you for providing such an extensive support structure and warm guidance and for making me feel like one of the gang as soon I arrived in Providence. Thanks also to fellow members of Phil's Contested Illnesses Research Group—especially Meghan Kallman, Elizabeth Hoover, Dave Ciplet, Tania Jenkins, Alyssa Cordner, Bindu Panikkar, Tyson Smith, and Mercedes Lyson, all of whom generously read iterations of this book. The chapters that follow are certainly the better for your advice. Kathy DeMaster, thank you for your laugh and for being an inspiring friend and next-door colleague in our old carriage house offices. Jeanne Loewenstein, thank you for making the Urban Environmental Laboratory feel comfy, even when I was a Providence newbie in awe of the humidity; and of course, thank you for reminding me to bring my keys.

I am grateful to Truman State University for the rigorous, unmatched undergraduate training I received there. Hena Ahmad, you were the first professor who made me believe that I might possibly be a competent researcher. You gave me the confidence to pursue my distant dreams and find my voice. Lloyd Pflueger, you taught me how to write—and how to make good writing sessions feel like long, deep meditations. This has made all the difference. Robert Graber and Elaine McDuff, you pushed me to develop my sociological imagination and theoretical muscles. I hope all of you know what inspiring people and teachers you are.

Thank you also to the Department of Sociology at Colorado State University for welcoming me into your ranks. It has been an honor and a blessing to work with such brilliant and warm colleagues, dedicated and accomplished graduate students, and environmentally minded undergraduates. I am grateful every day to work in a department where I learn and teach in personally and professionally fulfilling ways. I look forward to the years ahead.

On a personal note, I thank my friends and family, who have provided unwavering love and emotional support. You have made my life extraordinarily stable, even with three cross-country moves, job changes, and other adventures. Thank you especially to my husband, Matt Kazy, who has given so much for my career and this project. I am lucky to have such a fun, loving, smart-as-a-whip, and astute partner in life, travel, learning, hiking, dog raising, camping, and so much more. Thank you for being my best friend and a constructive critic and for bringing the warm and loving Kazy family —Joel, Melissa, Sarah, the boys—as well as all the Walkers and Sasses into my life. A special thank you to Barbara Walker for your support and for teaching me about Oak Ridge.

I thank my parents, Michael and Barbara, for empowering me to develop my own worldview—even when that meant raising a thirteen-year-old vegan and Bob Dylan devotee. You filled my childhood with security, love, laughter, and learning, a foundation whose solidity I couldn't fully appreciate until years later. I thank my sister, Genevieve, who let me drag her into the creek to catch frogs, "teach" her from my tiny chalkboard, and explore countless hiking trails on family road trips—and for bringing Brad and all his wit into our family. I thank my grandfather Henry, for teaching me how to sit still and feel wonder at the natural world; and my grandfather Peter, for teaching me how to speak truth to power. I thank my grandmother Virginia, for teaching me that education and travel are two experiences that no one can ever take from me; and my grandmother Lynne, for teaching me the beauty of good writing. I thank my aunt Michelle and my uncle Mark for taking me seriously, even when I was a six-year-old trying to hang out with the adults. And I am so grateful for my support system of friends, old and new. As I settle into Fort Collins, I miss many of you but feel blessed that you make so many other places and spaces also feel like home.

Finally, this book wouldn't exist without the graciousness and candor of people living in uranium communities across the Four Corners region. Fritz Pipkin, Barbara Pipkin, Steve, and all members of the Victims of Mill Tailings Exposure organization: you have been remarkable friends over the years, and I wish I could do more to help you in your daily fight. I am thrilled to be able to help you tell your story. To all other uranium community residents, thank you for letting me into your worlds and sharing your histories, perceptions, and colorful, funny stories. I have loved my time in the field and look forward to learning more from you in the years to come. To all the other generous people who let me explore your daily lives—members of social movements, specialists with public health agencies, uranium industry leaders and workers, and "fourth generationers"—thank you for your time, honesty, and willingness to discuss how these issues shape your lives.

The Price of Nuclear Power

1

Introduction

The Paradox of Uranium Production in a Neoliberal Era

It's a midwinter day on the Colorado Plateau; dry, whipping mountain winds add a chill to the sunshine. I am with Fritz and Barbara Pipkin, lifelong residents of rural Monticello, Utah, and leaders of Monticello's Victims of Mill Tailings Exposure, an environmental justice organization. Today the Pipkins and I are driving around Salt Lake City, attending various meetings with state public health officials and political aids. I listen to heated discussions about contested cancer clusters found among Monticello's 1,900 residents. In Fritz and Barbara's isolated, tight-knit town, cancer or chronic illness is a reality in nearly every household. While the federal government denies a connection, most residents link their diseases to uranium exposure from the Monticello Uranium Mill and two related Superfund sites.

Although the community's plight remains invisible to most contemporary Americans, Monticello played a key role in establishing the United States as an atomic superpower during World War II and the Cold War. The federal government-owned Monticello Uranium Mill, which operated adjacent to residential areas of town from 1942 through 1960, created the town's primary Superfund site.[1] The second site encompassed four square miles of the community, including residential and commercial buildings constructed by residents using radioactive remnants of uranium tailings piles from the primary site.[2] Though the federal government remediated both sites by the late 1990s, the environmental justice and health concerns of community members who were not directly employed by the industry remain largely unaddressed.[3]

Given this context, most people would expect Fritz and Barbara Pipkin to reject contemporary expansion of the uranium industry. But they do not. Despite spending years lobbying politicians for both healthcare and an acknowledgment of the industry's legacies in Monticello, the Pipkins tentatively support renewed uranium development.

Amid global concerns about climate change and fuel availability, nations such as China, India, and the United States are turning to nuclear power as part of an "all of the above" approach to energy policy. In the words of the World Nuclear Association, "increasing energy demand, plus concerns over climate change and dependence on overseas supplies of fossil fuels coincide to make the case for increased use of nuclear power," driving increased uranium production in communities of the Colorado Plateau.[4] Nuclear renaissance has ample political support and economic potential in rapidly developing contexts such as China. Efforts to define and fund nuclear power as a renewable energy source in the United States have increased as well and are equally motivated by substantial political-economic interests. For example, in 2009, Utah state legislator Aaron Tilton worked to have nuclear power officially defined and funded as renewable energy in the state; at that time, his company, Blue Castle Holdings, was seeking approval to build a nuclear reactor in the town of Green River, despite public outcry against the project.[5]

This is the fundamental paradox of renewed uranium development: the people and the communities that are most damaged by the legacy of uranium production are often constrained by historical and economic circumstances to support industry renewal. Such structural violence is brought into stark relief when we consider the persistent poverty of these communities. Further, because nuclear power's ability to provide socially sustainable energy remains in question (and because the legacies of U.S. uranium production remain chronically underaddressed), politicizing and expanding the industry as a source of renewable energy that is nonetheless characterized by policymakers as socially sustainable is ethically dubious. This is particularly true when such moves implicate uranium communities in consequent production expansion and its accompanying social dislocation, natural resource dependence, and potential environmental risk.

As I listen to Fritz and Barbara's ambivalent support for renewed uranium production, I cannot overcome my surprise and confusion. By this point in 2007, I have spent two years conducting fieldwork in southern Utah uranium communities. After witnessing so much suffering, I am shocked by even tentative support for industry renewal. Yet as I continue my fieldwork during the next four years, I will come to understand, even empathize with, the nuances of poverty, dependence on natural resource economies, and notions of environmental justice that may emerge when people with so little economic privilege are surrounded by land with such abundant mineral wealth.[6]

Still, the question persists. Why do the very people most affected by uranium's legacies even tentatively support nuclear renaissance? Why do some of them see industry expansion as a chance *for* environmental justice? After nearly twenty years of conducting health surveys, mapping cancer clusters street by street, and holding late-night meetings around their kitchen tables, Monticello's Victims of Mill Tailings Exposure are still fighting to gain federal recognition and community-wide compensation for the town's high rates of cancer, respiratory ailments, blood and reproductive disorders, and bankrupting medical bills—all of which residents link to long-term uranium exposure.[7] Monticello is no isolated case, and throughout this book I will show that similar environmental justice and health outcomes have emerged in many other communities affected by Cold War–era nuclear technologies.

The summer before our trip to Salt Lake City, I had visited with the Pipkins in their living room, a space peppered with reminders of their activism: a carefully crafted six-by-six-foot map of cancer clusters in Monticello, pictures of community children lost to leukemia, a work station for organizing temporary cancer screening clinics. Barbara explained, "Listen, we realize how strange it sounds. But we live in a poor place built on uranium, a place that needs it. And we have to have faith that regulations are better. We're just this little dot in the middle of nowhere. How could we say no?"

Uranium communities such as Monticello must negotiate between historical legacies and contemporary energy development, between environmental health issues and economic justice, between spatial isolation and global energy markets. What conditions lead to mobilized sites of acceptance or sites of resistance to uranium production's renewal? What opportunities and constraints feed activism and shape its goals? What do conflicts reveal about nuclear energy's social sustainability as well as shifts in the meanings of environmental justice in the United States and other neoliberal political economies? To illustrate: recently, and controversially, Energy Fuels Resources has emerged as a significant driver behind nuclear renaissance on the Colorado Plateau. The corporation's proposed Piñon Ridge Uranium Mill has been zoned into Paradox Valley, an agricultural pocket in southwestern Colorado hugged by red-rock mesas. The decision will be politically significant because Piñon Ridge is the first U.S. uranium mill to receive a permit since the end of the Cold War. Further, Energy Fuels Resources' recent merger with Denison Mines has made the corporation the largest, most powerful uranium producer in the United States. Reactions to the mill continue to unfold, a signal of the contentious notions surrounding the nation's sustainable energy development and our ideas about environmental justice.

The situation extends beyond one permitted mill in a rural pocket of the United States. It represents general patterns in energy policy, as nations scramble to address worldwide climate change within a complex nexus of technological innovations and globalized markets. Similar patterns of vulnerability and divergent activism accompany hydraulic fracturing, coal mining, and offshore oil and gas drilling—all of them mobilizing in spaces where communities and energy production intersect. Energy policy is at a critical historical juncture, one that demands an ethic of social justice toward the communities that provide its raw materials. Yet the residents themselves express varying notions of justice, illustrating the impact of hegemonic neoliberal ideologies, policy approaches, and subject formation on notions of environmental justice and expressions of social activism.[8]

These shifts in uranium communities and what they signal about energy development and transformative social change are my focus in this book. By way of the histories and narratives of uranium community residents, past and present, I demonstrate that framing nuclear energy as a socially sustainable energy source is ethically problematic and sociologically unreliable. The industry's environmental justice and health legacies continue to plague persistently poor and isolated communities in the Southwest, even as corporations and political allies assure them that nuclear renaissance will bring economic revitalization and energy innovation. But these claims remain suspect, given the industry's historic economic volatility, the massive expense of current projects and repairs, public discomfort with nuclear power since Japan's Fukushima tragedy, and the potential effects on uranium communities that struggle with the structural constraints of persistent poverty, spatial isolation, and natural resource dependence.

PHOTO 1 Paradox Valley, home of the proposed Piñon Ridge Uranium Mill, epitomizes the vast spatial isolation of the Four Corners region. (Photo by Matthew Kazy)

Nuclear Renaissance Comes to Life

Today, several years after my conversations began with Fritz and Barbara about renewed uranium production, industry renewal is no longer an abstraction but a reality. In 2011, the government gave Energy Fuels Resources initial permits for the construction of Piñon Ridge Uranium Mill, and thus far all legal challenges been have surmounted. The corporation still plans to build the mill in Paradox Valley, Colorado, just a few miles from the towns of Nucla and Naturita. Although the permit remains contested and uranium markets are unstable, people in these nearby communities have overwhelmingly accepted Piñon Ridge's presence, making the mill a significant and controversial symbol of renewed uranium production and nuclear renaissance.[9]

To investigate the newly permitted mill and the vociferous community response it had created, one hot summer day I found myself driving down an empty two-lane state highway in southwestern Colorado. In this hostile, isolated region, a few mining communities still survive, but others have become contaminated ghost towns. As I traveled from Telluride's ski-town luxury into Nucla's and Naturita's rough and rugged sparseness, I bisected family ranches and green grasslands before descending into an arid landscape dominated by red-yellow dust, flat-top mesas, curving canyons, and surprising pops of green where irrigation canals sustain life. Within a mere sixty miles, I had visited two distinct worlds, one privileged and one impoverished, though each depends on natural resources for economic survival.

Telluride's crowded main street overflows with upscale clothing and recreational equipment boutiques, vegan and organic eateries, and tourists in Cloudveil ski gear driving luxury sedans and SUVs. A free gondola lifts happy vacationers to their second or third multimillion-dollar mountainside homes. Oxygen bars and microbreweries cater to skiers, snowboarders, and the occasional celebrity. On Naturita's modest Main Street, several storefronts sit vacant, a few bars and small diners compete for steady business, and rusting pickups meander along empty roads. Social dislocation—the patterned experiences of instability and powerlessness among residents of rapidly changing market-based economies—permeates the daily lives of community members.[10] Bungalows, doublewide trailers, and modest single-family homes line a few in-town streets, but paved roads give way to gravel almost as soon as I turn my car off Main Street. Although the sister cities of Nucla and Naturita share water treatment services and other infrastructure to conserve public spending, the local governments remain resource-poor. A for-sale sign hangs on the local middle school; reduced funding and enrollment numbers have forced the towns to consolidate students into one building.

Cheerful gardens and mowed lawns do not disguise the social dislocation that plagues most of the households. Fewer than 50 percent of the children in Nucla and Naturita live in two-parent homes, and nearly 20 percent receive special education services. Approved applications show that close to 60 percent of local school-aged children depend on free lunch programs, and recent figures indicate that the high school dropout rate in Nucla ranges from 10 to 27 percent. Older students and other residents contend with rampant drug use, and the communities have growing reputations for crystal methamphetamine use and manufacture.[11]

Many community residents, particularly those who describe themselves as 'fourth generation', are strongly in favor of the mill, largely due to the employment opportunities they believe it will provide. But their support extends beyond basic economic identities; personal and community identification with uranium, attachment to place, and a sense of patriotism also motivate their acceptance. As I visited with Naturita's bar patrons, grocery store customers, and diner employees, my casual questions uncovered stronger support for renewed uranium production than I'd heard from Fritz, Barbara, and other Monticello residents. After the bust of the second U.S. uranium boom in the early 1980s, Nucla and Naturita found themselves dealing with chronic recession, environmental health problems, and even more spatial isolation than Monticello. Nonetheless, support for industry renewal has mobilized residents' social activism and dominates their built environment. When I walked into the visitor center's simple prefab building, my eyes were immediately drawn to a neon orange sign high on the wall: "Yes to the Mill!" As I talked with the center's employees, I heard one after another speak in favor of the Piñon Ridge Mill and identify with the uranium industry as part of their communities. But even as residents' historical and economic realities were mobilizing community support, they were continuing to marginalize the communities, constrain economic opportunities, and limit people's ability to challenge the effects of the privatized industry.

Structural Violence in America's Sacrifice Zones

Renewed uranium production in the United States remains invisible to most Americans.[12] This is understandable: the vast Colorado Plateau is cloaked in desert mystery and has long been a sacrifice zone for American atomic ambitions.[13] The plateau covers 140,000 square miles in the Four Corners region, a landscape defined by its stark beauty and sparse human population. Yet the region is uranium-rich, so the area is continually affected by global nuclear trends.[14] Today, even after the Fukushima disaster, nuclear power retains its image as a climate-friendly energy source, and increasing demand for uranium from developing countries such as China is pushing global markets to expand.[15] Even the United States, notorious for nuclear inertia since Three Mile Island's near-meltdown in the late 1970s, has scheduled the construction of two new nuclear reactors in Georgia and granted permits for many others around the country.[16]

As Monticello illustrates, lingering environmental, health, and social impacts from two previous uranium booms have never been adequately addressed. But this history has not halted renewal, as the Piñon Ridge

Uranium Mill's permit proves.[17] In this sparse pocket of the American West, the mill and conflict over it have become symbols of the tension between contemporary energy development and uranium's local legacies—where structural violence complicates and shapes regional class conflict.[18] While these legacies marginalize communities such as Monticello, Nucla, and Naturita, they privilege the global drivers of uranium markets. Uranium communities have limited economic and energy development opportunities, and resisting renewed production is challenging when residents deal with daily limitations of poverty. Meanwhile, a handful of multinational corporations such as Energy Fuels Resources control the industry's renewal and encourage rapid market growth by buying out mines and leftover infrastructure from previous booms. These companies also fuel growth and shape preliminary royalty agreements with the holders of mining claims. Between 2002 and 2007, prices for ore and refined uranium skyrocketed from below ten dollars per pound to more than ninety dollars per pound, prompting speculation and prospecting.[19] In 2009 alone, western companies and wealthy residents staked more than 32,000 uranium claims. Though prices have recently fallen to between forty-one and forty-five dollars per pound, economic incentives are likely to increase as Chinese markets and a 54-billion-dollar stimulus for new U.S. nuclear reactors fuel demand.[20] Yet these structural constraints on individual agency remain invisible to the residents of Nucla and Naturita, obscured by decades of identification with resource-based industries as well as the hegemonic power of neoliberal ideology.

Divergent Mobilization: Sites of Resistance, Sites of Acceptance

After decades of structural violence related to chronic poverty and natural resource dependence, uranium communities have become fertile ground for divergent modes of social activism—a collision of varied notions of environmental justice, land use, and economic development. When the Colorado Department of Public Health and Environment announced initial approval for the Piñon Ridge Mill in 2011, social tensions erupted.[21] These tensions have intensified after several court battles, as the health department was required to reexamine its decisions and reissue mill permits. Communities and neighbors are divided; death threats have even been exchanged. "We need the mill bad. . . . We're just hanging on by a thread," my new visitor's center friend told me. "We've been involved with uranium for years. We know it better than other people do, *especially* those folks in Telluride, and it's safer now. The regulations are better and Energy Fuels will treat us fine because George [former CEO of Energy Fuels] is a local." Throughout Nucla, Naturita, and other spaces that nurture what I refer to as *sites of acceptance*, support for renewed uranium production thrives. Citizens in these sites believe that natural environments must be used for industrialized production to meet society's energy needs, and they accept risks for the sake of the potential rewards, economic and otherwise, that development may provide. Unlike the boosterism in former company towns or community economic identity seen in Appalachian coal country, citizen attitudes in sites of acceptance have a multifaceted, historically contingent complexity.[22]

Opponents to the Piñon Ridge Uranium Mill are primarily based in Telluride and Paradox, with others scattered across the region. Together, they create what I call *sites of resistance* to renewed uranium production. These citizens have mobilized against industrialized production practices and risks, often using the precautionary principle as a guide. Some activists align with formal organizations, while others express opposition more quietly. When I spoke with Heather, director of the Telluride-based anti-industry organization Sheep Mountain Alliance, she said something quite representative of activists' perceptions in sites of resistance: "These people who say 'Regulations are much better these days,' well, they aren't. . . . So saying, 'The state is going to protect us,' well, no, they are not actually, and we need to continue to fight for those protections."[23]

But instead of banding together to fight for protections, activists have turned on one another. Class tensions are a central point of conflict. For example, Sheep Mountain Alliance inflamed Nucla residents by flying the neo-hippie jam band Phish into Telluride to perform an "anti-radiation, anti-mill concert" in summer 2010. In other words, divergent regional activism illustrates the toxic ambivalence within the nuclear renaissance. Renewal visibly intersects with legacies of environmental injustice, persistent poverty, and spatial isolation; activism splinters along class lines to create sites of resistance and acceptance. Single sites of acceptance typically emerge as social movement organizations; and in combination, these single organizations or informal groups help power what I call the *triple movement*, in which markets for commodities such as uranium become part of community social fabrics and are defended and supported by people as part of local culture and norms, despite the historic instability of those commodity markets.

Social sustainability remains elusive in uranium communities as their residents diverge around renewed production. During nearly a decade of fieldwork, I have learned that complexity and tension define uranium production's social impacts. Although I entered the field thinking that support for renewed production was counterintuitive, I left realizing that all-or-nothing responses tend to remarginalize people and communities that are already experiencing deep social dislocation. Yet the social tension in uranium communities has been far from productive or empowering for residents. Rather than driving progressive or transformative social change, tensions harden over time and create impasses to sustainable and cooperative regional development. Rather than collaborating on regional energy development initiatives (such as solar power installations, for example), activists and communities fight one another instead of their less-visible structural constraints. These struggles remain largely invisible in U.S. renewable energy debates, where nuclear waste storage and power plant safety grab the headlines.

Throughout this book, I will show how the nuclear renaissance generates and exacerbates structural violence—enhancing social divisions, exacerbating inequalities, and threatening community-level social cohesion. Renewed uranium production is not a socially sustainable form of energy development in rural communities because it creates few opportunities for rural residents to access sustainable livelihoods not completely dependent on unstable global markets. Without socially sustainable systems for uranium communities, residents have little ability "to cope with and recover from stresses and shocks, to adapt to and exploit changes in its physical, social, and economic environment, and to maintain and enhance capabilities for future generations."[24]

Why does social sustainability matter? Renewable energy carries strong connotations of social and environmental sustainability, vital considerations as the United States considers nuclear power's role in our energy ambitions. When former Environmental Protection Agency administrator Lisa Jackson took office in 2009, she asserted that renewable energy should act as a social panacea, "cutting through a thicket of thorny social ills and solv[ing] long-standing problems across the entire spectrum of American life" such as environmental, social, and economic insecurities.

Renewable energy carries these connotations internationally as well, with groups like the United Nations characterizing renewable energy as socially sustainable.[25] My evidence, however, suggests that it is disingenuous to fund nuclear power as a socially sustainable, renewable approach to energy policy. Instead, uranium communities contend with a complex, unsustainable web of persistent poverty, enduring health problems, conflicting conceptions of environmental justice that split social activism, and natural resource dependence—each unfolding in a neoliberalized political-economic context in which privileging free market systems has become normal.

Major Players and Places

In this book, I introduce spaces, people, and places central to the nuclear renaissance yet invisible to most energy consumers. Among them are several uranium communities that have become battlegrounds and where sites of acceptance and resistance mobilize (see map 1). Nucla is a Colorado town of 732 people, with 22.1 percent of its population living below the poverty line. Founded in 1893 by small group from Denver, the town was an intentional community founded on progressive social ideals and focused on communal living. Naturita, located just a few miles from Nucla, is a town of 635 people, with 30 percent of the population living below the poverty line.[26] Not far away are the unincorporated communities of Paradox and Bedrock, both in Paradox Valley.[27] They all have primarily agricultural histories, though their economies also depend on extractive and energy-based activities such as coal mining, uranium mining and milling (when markets are booming), and a local coal-fired power plant.[28] All of these communities have weathered persistent poverty and chronic recession since uranium's last bust in the early 1980s.

Monticello, located about sixty miles west of Nucla and Naturita, is a rural uranium community of 1,958 people in Utah's sparsely populated San Juan County.[29] The town lies about sixty miles south of Moab, just east of Canyonlands National Park's sprawling red rock landscape. The high-elevation community enjoys crisp air, more than three hundred days of sunshine every year, and a beautiful high desert landscape in the shadow of the Abajo Mountains. Yet despite its natural wealth, San Juan County is among the poorest in the state, with almost 30 percent of the county's

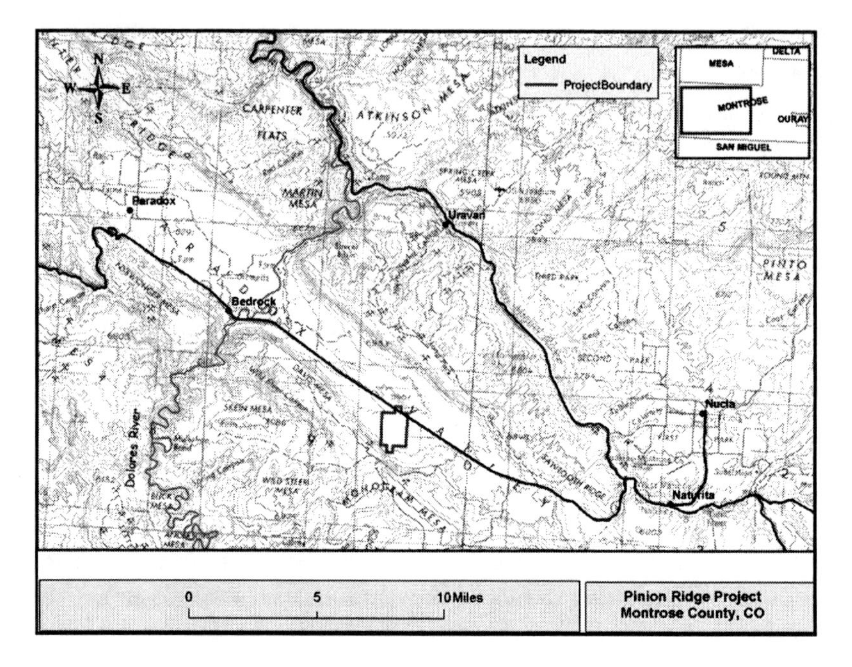

MAP 1 The Piñon Ridge Uranium Mill's proposed site, outlined in the center of the map. *Source:* Colorado Department of Public Health and Environment, 2014.

population living below the poverty line.[30] Monticello has long experienced economic boom-and-bust cycles due to community dependence on natural resource–based production, and it still contends with cancer clusters related to the legacies of uranium production, though its economy is more diversified now.

Telluride, Colorado, is a central site of resistance to the Piñon Ridge Mill. The community has its own mining legacy; before the 1970s, it was a silver mining camp and a center of gold, zinc, copper, and lead mining. Telluride has a history of boom-and-bust cycles and was nearly a ghost town by the 1970s. But skiers revitalized the area, capitalizing on the town's towering peaks, waterfalls, and forests and establishing the small community as a world-renowned ski resort and luxury community. Today, the 2,000-person town is better known for its box canyon beauty, ski slopes, and multimillion-dollar homes than for its mining history.[31] Nonetheless, Telluride has been affected by the U.S. recession since 2008, with home prices falling, construction halting, and the poverty rate doubling; today 19.9 percent of households live below the poverty line.[32] As poverty increases, the town fiercely guards its tourism-based economy and its image as a pristine ski haven.

Energy Fuels Resources and the Colorado Department of Public Health and Environment are the key institutions shaping renewed uranium production on the Colorado Plateau. Energy Fuels, which is a Toronto-based, publicly traded uranium and vanadium resource firm, owns Piñon Ridge Mill's proposed site and the permits for its construction. Despite its place on the Canadian stock exchange, Energy Fuels focuses its activities on land in the United States—principally the Colorado Plateau. Importantly, former CEO George Glasier, who retired after the corporation received county-level approval for the mill, owns and lives on a ranch in Nucla, has acted as a pro bono lawyer for residents and the town, and, as I will show, is trusted as a local by people in these communities. To highlight its regional ties, Energy Fuels has established offices in Nucla and Lakewood, Colorado, as well as in Kanab, Utah. During public meetings and on the company and mill websites, the corporation emphasizes that most of its management team is comprised of people who have lived in Colorado for many years.[33]

Since announcing its plans and beginning preliminary licensure steps for the Piñon Ridge Mill in 2007, Energy Fuels has made a concerted effort to expand uranium mining operations in the region, enlist support from local residents, and fulfill the Colorado Department of Public Health and Environment's licensure requirements. Its acquisition of Denison Mines' U.S. assets in summer 2012 allowed the corporation to acquire the White Mesa Mill (the only operational uranium mill in the United States) and many uranium mining claims. Now the nation's largest conventional uranium producer, Energy Fuels continues to renovate uranium mines around the proposed mill site and its new holdings, which would provide about 70 percent of the uranium processed at Piñon Ridge.[34] The corporation also holds the mineral rights to large swaths of uranium-rich land in Colorado, Arizona, and Utah, which it intends to explore (see map 2).[35]

The Colorado Department of Public Health and Environment oversees and regulates almost all uranium- and nuclear-related activity in the state as per its Agreement State status with the U.S. Nuclear Regulatory Commission.[36] The department's Radiation Management Unit handles most of the applications and permitting processes related to the Piñon Ridge Mill. It has held multiple public meetings, sometimes in concert with Energy Fuels Resources, to address public concerns about the facility

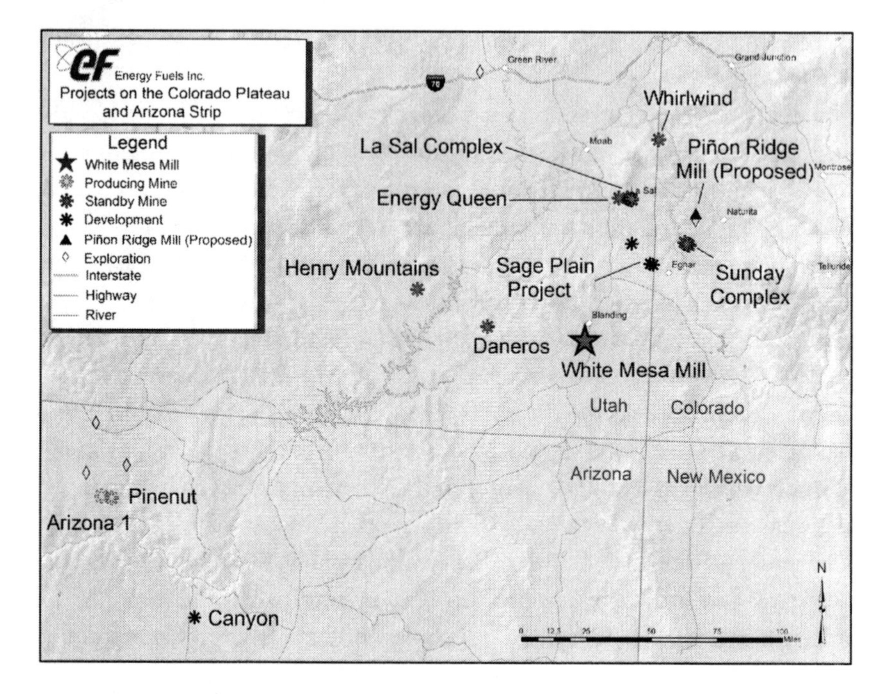

MAP 2 Energy Fuels' uranium milling and mining projects in the Four Corners region. *Source:* Reproduced with permission of Energy Fuels Resources, Inc.

and renewed uranium production more generally. Though the department acts as the primary regulator of uranium production in Colorado, it has faced mounting budget cuts and financial constraints as the recession has affected public spending.

Several social movement organizations have mobilized. In Nucla and Naturita, sites of acceptance for the Piñon Ridge Mill mobilize through shared community spaces such as the visitor's center; through media sources such as the local newspaper, the *San Miguel Basin Forum*; through prominent public figures and trusted residents; and during public meetings. More professionalized organizations have mobilized sites of resistance to the mill and renewed uranium production, forming a coalition of activists and social movement organizations that include the Sheep Mountain Alliance and the Paradox Valley Sustainability Association. The most prominent of these groups, the Sheep Mountain Alliance, was formed in 1988 as a grassroots organization working to protect Telluride's natural environment for common use by all people, regardless of social class. It has filed several lawsuits against both Energy Fuels and the

Colorado Department of Public Health and Environment, claiming that permitting processes for the mill were inadequate. Other organizations, such as the Paradox Valley Sustainability Association, the Western Colorado Congress, and Grand Valley Peace and Justice, have long lobbied the department to reject Energy Fuels' licensure application.[37] In contrast, activist groups such as the Victims of Mill Tailings Exposure and Uranium Watch, both located in Utah, advocate for cautious approaches to uranium renewal and focus on addressing industry legacies and regulatory compliance rather than fighting against renewal.

The Sociological "So What?": Why Renewed Uranium Development Matters

The story of uranium development in the United States is a captivating tale—complete with wars and bombs and prospecting in the rugged American West. Sociologically, the industry's legacies and potential renewal illuminate important patterns in how power and privilege are distributed in societies in which neoliberal ideologies have become hegemonic, how this hegemony affects people's notions of environmental justice, and how, in turn, goals of activism may fail to initiate transformative change that addresses persistent poverty as a component of sustainable energy production.

The Neoliberal Turn in the U.S. Political-Economic Landscape

Neoliberalism describes a set of ideologies that have been enacted in the United States since the early 1980s. These ideologies frame free market capitalism as better than other socioeconomic systems and privilege market-based logic as superior even in noneconomic contexts.[38] Neoliberalism is expressed through policy discourses such as privatization, marketization, state deregulation, market-friendly reregulation, and the creation of self-governing individuals.[39] *Neoliberalization* describes the process of implementing policy measures such as free trade agreements and devolving federal governance to cash-poor states and localities.[40] While neoliberalism has been "fetishized . . . as a single, monolithic and undifferentiated process," it can be best understood as a "diverse and interlinked set of practices that reflects a heightened, evolved and more destructive form of capitalism."[41] Neoliberal ideologies are variously deployed, interact with different people in various places and at multiple scales, and thus manifest in widely divergent ways.[42]

Neoliberalism began to dominate U.S. and western economic policy in the 1980s, when policymakers implemented *rollback neoliberalism* by actively shrinking the federal government's funding and its ability to provide social programs and resources, which exacerbated national inequality and poverty.[43] In response to public protest and pressure, *rollout neoliberalism* began in the 1990s, rebuilding the federal state to accomplish "aggressive reregulation, disciplining, and containment of those marginalized or dispossessed by [rollback] neoliberalism."[44] Nonetheless, problems related to poverty and inequality continue today, even though rollout neoliberalism created more public buy-in and thus helped make neoliberal ideology normative and hegemonic.

Devolved governance—that is, assigning leadership responsibilities to smaller levels of government such as municipalities—has remained part of the new neoliberal normal in the rollout phase. Yet small levels of government often have fewer financial resources, even with their increased governance responsibilities. Consequently, communities often seek private funding to support public services and may encourage local corporate investment, such as energy development, as they become increasingly responsible for governance. Neoliberal policies continue to impact uranium production and communities. Industry privatization, de- and reregulation of related governance mechanisms, and corporate self-monitoring are just some of the neoliberalized traits of the uranium industry.[45] For example, state commodification of uranium in the 1940s and 1950s allowed small-scale prospectors to collect government bonuses and gain economic security and relative independence. However, contemporary enclosure and corporate ownership of uranium deposits, which began in the late 1960s, ensured uranium's privatization. As deposits were enclosed, residents were excluded and have become increasingly dependent on unstable wage labor markets.

Although structural relations are key, neoliberalism's power derives from its perceived legitimacy. People throughout western societies think that free-market privilege is completely normal; they intuitively connect neoliberalism with historically American notions of individual freedom even as they hone a correspondingly narrow, individualized sense of ethics and responsibility to others.[46] This logic wields enormous hegemonic power and influence in creating a new normality by way of neoliberal governmentality.[47] The state, practicing a new "art of government," constantly intervenes in creating structures that facilitate free markets.[48] Decisions filter through cost-benefit frameworks, normalizing market privilege and the effects of development.

The neoliberalized state "conducts conduct," helping citizens internalize norms of market-based logic that in turn encourage them to prioritize economic development. People "self-regulate their behavior in ways consistent with neoliberal logic" in the context of natural resource extraction such as uranium development or, as research suggests, other energy production, water privatization, and pollution.[49] As the state retreats in de- and reregulated contexts, people become more atomized, or self-governing, particularly in rural uranium communities where spatial isolation and peripheral status already create distance from urban centers. Further, "opponents of [development] are discursively framed as irrational and unwilling to absorb necessary costs that would benefit their neighbors and the nation as a whole."[50]

There are examples of neoliberalism's normalization all around us. In the United States, the United Kingdom, Western Europe, Australia, and other neoliberalized contexts, citizens think of themselves as citizen-consumers who vote with their currency. Because personal retirement accounts and mortgages are increasingly implicated in the market system, we stay keenly aware of daily stock prices and market dynamics. We begin to conduct cost-benefit analyses in noneconomic situations in daily life, such as how to schedule our days, raise our children, or plan our daily meals. Market-based logic becomes increasingly normal as a guide for life. Yet neoliberalism's hegemonic power also helps to normalize environmental degradation, in some instances fundamentally reshaping activism and its goals.

The Role of Natural Resource Dependence

With uranium markets increasingly neoliberalized, uranium communities are especially vulnerable to boom-and-bust cycles.[51] Because local resource economies are embedded in globalized commodity chains and natural resource–based market systems, residents' social dislocation often takes the form of economic instability as volatile markets leave them struggling to feed their families and keep their households afloat.[52] Why do unstable global natural resource markets and commodity prices lead to higher-than-average rates of un- or underemployment in towns already limited by spatial isolation and poverty? While global prices for natural resources tend to fall, costs of local production rise over time, creating community economic systems vulnerable to shocks and households powerless against chronic poverty.[53] Dependence persists in some rural communities despite repeated experiences with boom-and-bust cycles, signaling historically and economically generated forms of structural violence. Rapid booms may challenge community infrastructure such as roads, plumbing, and schools; lead to intense conflicts between newcomers and old-timers; and correlate with increased crime, traffic, cost of living, and general social disruption. Though disruptions may lessen over time, continued dependence on natural resource markets can perpetuate social dislocation among individuals and social disruption within communities.[54]

High rates of persistent poverty plague rural, natural resource–dependent communities such as those on the Colorado Plateau.[55] Rural deindustrialization, extralocal corporations' power over communities, and core-periphery relations between urban and rural areas act as key mechanisms in U.S. rural impoverishment in resource-dependent areas.[56] Rural distancing, a practice by which spatially isolated communities are treated as peripheral regions, is another key mechanism.[57] Thus, Colorado Plateau uranium communities can be characterized as disempowered places or peripheral regions that "in a market economy will experience economic rise or decline in response to such circumstances as demographic changes, technological advances, and the depletion of resources."[58] In these spaces, uneven development creates conditions in which "social inequality [is] blazoned into the geographical landscape."[59]

In uranium-dependent regions, this inequality is tangible; "state-sanctioned sacrifice zones" in the American Southwest have superimposed "invisible nuclear landscapes" on regions still occupied by Native Americans and impoverished populations. Isolated settlements in the region, especially uranium communities, were deeply affected by atomic age development and show evidence of "nuclear colonialism."[60] As I will discuss in chapter 2, booms and busts in these communities have highlighted their peripheral positions in relation to more politically and economically powerful core areas.[61]

Still, natural resource dependent areas might transition into a non-extractive dependence on resources; high-amenity communities might host tourism and recreation facilities or market themselves as destinations for second-home owners.[62] For example, Moab transitioned from "uranium capital of the world" to a mountain biking and national park mecca. Telluride also successfully transitioned from a mining boomtown to playground for wealthy skiers. While this strategy may lead to higher employment rates, higher property and home values, population growth, and new side businesses, costs can include economic vulnerability to downturns and the creation of low-wage service-sector jobs, which are typically part time or seasonal and offer few benefits.[63] Apparently, tourism is no more likely than typical extraction-based economies to secure significant economic sustainability.

In short, boom-and-bust volatility, persistent poverty, peripheral status, and spatial isolation combine to limit economic development options in spaces such as uranium communities. After decades of empirical research, scholars have found that natural resource dependence harms rural communities, specifically in its correlation with persistent poverty.[64] Because they are dependent on extralocal market demands, community economies and social fabrics are ravaged when demand falls.

Alternative Notions of Environmental Justice and Sites of Acceptance

When rampant social dislocation results from persistent poverty and is further structured by spatial isolation and natural resource dependence in a neoliberal economy, different notions of environmental justice emerge through mobilized sites of acceptance and sites of resistance. These notions inspire types of mobilizations whose tactics and goals differ from typical environmental justice activism in the United States—outcomes that signal important shifts in how justice is conceptualized where social inequality is 'blazoned on the landscape,' where neoliberalism is normalized, and where the transformative potential of contemporary grassroots activism cannot be the ultimate focus of activists.

Environmental justice is a contested term with various connotations. For some, it means that "all people and communities are entitled to equal protection of environmental and public health laws and regulations."[65] For others, it means that all people have a right to feel safe where they live, work, and play. The term also refers to various disenfranchised groups establishing procedural justice by acquiring local autonomy over land use decisions, especially through appeals to procedural equity.[66] Conceptions of environmental justice have expanded since the 1980s, even among academics, who now articulate the provocative notion that "the issues involved are about how, exactly, we are immersed in the environment, and the manipulation of nature around us."[67] Environmental justice has "moved from . . . a reflection of social injustice generally to being a statement about the crucial nature between environment and the provision of justice itself, . . . about the material relationships between human disadvantage and vulnerability and the condition of the environment and natural world in which that experience is immersed."[68] In other words, especially in cases of natural resource dependence, material interactions between society and landscape help structure poverty and social dislocation while shaping varied notions of justice among activists. Divergent notions of environmental justice have emerged as uranium community residents have faced varied forms and degrees of social dislocation—that is, as they have experienced powerlessness or vulnerability related to environmental degradation and to more immediate economic constraints as participants in free market systems.

The U.S. environmental justice movement began in the early 1980s, when activists created some of the first sites of resistance to toxic contamination (Love Canal, New York) and environmental racism (Warren County, North Carolina), which they linked to specific hazardous facilities in their communities.[69] Foundational studies revealed that higher percentages of people of color lived near hazardous sites in various study regions; and they later established that impoverished populations, regardless of race, were also disproportionately affected by industrial waste sites and pollution.[70] Over time, activists developed broader concerns, and *environmental justice* became a more inclusive term that took into account people's interactions with hazards and toxins in various environments and locations.

Yet even after community-based case histories and comparative analyses of activism became central modes of analysis, scholars remained focused on successful movements that served as sites of resistance, even in company towns. [71] Case studies examined environmental justice movements in a variety of military-industrial circumstances, including toxic and chemical pollution in the South's Cancer Alley area, military and hazardous waste left in rural (and sometimes Native American) communities, and health outcomes among populations exposed to pesticides.[72] From this research, a clear pattern emerged: in the United States, members of low-income communities, racial and ethnic minority groups, and indigenous groups "confront a higher burden of environmental exposure from air, water, and soil pollution from industrialization, militarization, and consumer practices."[73]

As characterized by sociologists, then, environmental justice activists target the federal government or industries to fight for increased environmental protections or increased precaution in policies on chemical use, land use, energy development, and other environmental measures. They mobilize sites of resistance to disproportionate exposure of poor, minority, or marginalized communities to industrial pollutants, potential health impacts, and lack of local control over land use.[74] Only select analyses discuss environmental justice outcomes explicitly in terms of neoliberalism's impacts on activism, and those that do focus on sites of resistance to its environmental costs.[75] Consequently, environmental justice activism is often romanticized as motivating creation of unequivocal sites of resistance, and case studies focus largely on successful movements organized by progressive activists who define environmental justice in uniform and fairly radical ways.[76] Messier movements related to land use or procedural equity are not as frequently analyzed.[77] Even top theorists tend to romanticize environmental justice activism and focus on notions that conform to disciplinary tropes. For example, Dorceta E. Taylor asserts that the movement acts as a powerful master frame for activists and observes that environmental degradation must have strong local salience for residents to mobilize around it.[78] Indeed, activists' notions of justice depend on community contexts. Yet the framework is not readily applicable to cases in which people's ideas of environmental justice may be shaped by neoliberal logic. Sociologist David Pellow's *Garbage Wars*, which examines Chicago's Robbins neighborhood and residents' active recruitment of a waste incinerator, provides a rare analysis of a site of acceptance.[79] Though Pellow observes different notions of environmental justice at work, his discussion does not offer rigorous conclusions that are applicable to other cases.

Environmental justice activism has also focused on "health effects caused by toxic substances in people's immediate or proximate surroundings."[80] Health social movements act as key sites of resistance in which activists frequently employ popular epidemiology (the use of lay science and surveys) and create citizen-science alliances to legitimate their observations about connections between industrial toxins and human health degradation.[81] While this framework offers a useful set of criteria for identifying progressive environmental justice mobilization, activism in *support* of potentially hazardous facilities or *denying* verifiable health impacts related to industrial pollutants remain understudied. We therefore know comparatively little about activism that creates distinct sites of acceptance for industrial production and its risks—that is, where activists define environmental justice as local sovereignty over land use decisions because of the unique ways in which their livelihoods and communities are embedded in those landscapes.

While many instances of environmental activism create sites of resistance to neoliberal and industrial development, some activists internalize and use neoliberal norms as they mobilize sites of acceptance.[82] For example, community forestry sites throughout the United States have created "hybrids between neoliberalism and . . . natural resource management."[83] Likewise, some First Nation Alaskans have adapted neoliberal worldviews as they operate their fisheries.[84] In Ecuadorian old-growth rainforests, cultural practices and neoliberalization processes have overlapped with indigenous identities to accommodate market-based logic.[85] Impoverished small-scale Chilean farmers have adapted to neoliberal policies in response to increasing inequality.[86] The U.S. Environmental Protection Agency's approach to hazardous waste treatment suggests that even environmental justice advocates use market-based logics, and California policies that encourage healthier eating are contributing to "neoliberal subject formation."[87] In short, neoliberal logic is helping to legitimate a new normality and new notions of environmental justice, which in turn nurture sites of acceptance for industrialization, privatized natural resource extraction, and environmental degradation.

Despite these varied notions of environmental justice, residents in uranium communities remain concerned about air and water contamination, long-term exposure to such contamination, and the overall degradation of nearby ecosystems.[88] These injustices continue to generate concern as residents such as Fritz and Barbara Pipkin identify cancer clusters and connect them to uranium exposure. Assessing the possible outcomes of nuclear renaissance, Tufts University public health specialist Doug Brugge warns that "the need to control and limit exposure to uranium seems more important than ever. If a new 'uranium boom' is indeed emerging, then efforts must be vigorous to limit environmental contamination and exposure." Yet given uranium communities' persistent poverty and economic constraints, some activists eschew concerns about environmental health outcomes. Unlike other groups, which have organized into sites of resistance around more radical notions of environmental justice and see uranium markets as socially dislocating, these activists envision globalized uranium markets as part of their communities' social fabric and their own identities. In their eyes, uranium markets are socially affirming, and local control over land use is the ultimate form of environmental justice.

Transformative Social Activism?

Neoliberalism's hegemonic status in the United States encourages such starkly market-based notions of environmental justice. While they make sense to many residents of Nucla, Naturita, and communities in similar situations, these notions are not transformative in their tactics, targets, goals, or ability to create meaningful social change. As neoliberalism has become normalized and internalized across societies in recent decades, scholars have been examining the transformative potential of social movements—that is, their ability to re-embed markets in social contexts or regulatory protections.[89] This potential is altered by new movement targets and logics. In cases of commodity production such as fair trade coffee or forestry certification, corporations and even third-party regulatory systems are able to co-opt and dilute social movements, compromising their power to enact fundamental changes to social structures and power differentials.[90] These findings signal shifts in what is sometimes called *social movement ecology*: now mobilized groups may have goals and tactics that do not lead to fundamental social change.[91] Activists may indeed be co-opted by corporations as "the ability of the corporate target to bring the interests of a challenging group into alignment with its own goals" has become extraordinarily well developed in neoliberalized settings such as the United States.[92]

Structurally, new corporate targets are symptomatic of a shrinking state encouraged by neoliberal policies that devolve governance to smaller units. Operating in a "weakened state," movements must address nonstate entities such as corporations and nongovernmental organizations.[93] Yet, this does not necessarily make these social movements "significantly more radical and disruptive" than movements targeting the state, as some have claimed.[94] Instead, as analysts of the radical Right to the City Alliance have observed, "transformative organizing recruits masses of people to fight militantly for immediate concrete demands . . . but always part of a larger strategy to change structural conditions in the world. . . . Transformative organizing works to transform the system, transform the consciousness of the people being organized, and in the process transform the consciousness of the organizer."[95] It remains sociologically vital, then, to examine how and when environmental justice activists may use neoliberal logic as their logic for social activism and to interrogate whether this dilutes the transformative potential of those organizations. Activism's transformative potential matters, especially in cases of environmental injustice, because grassroots activism provides one of the few avenues available for disempowered, marginalized communities to change inequitable social structures.

What Happens When Markets Become Part of a Community's Social Fabric?

Similar outcomes and mobilized sites of acceptance are visible in many energy production activities—not only uranium production but also hydraulic fracturing for natural gas and coal mining—and in economic sectors such as banking and finance. In other words, the patterns in social mobilization that I link to uranium production signal significant shifts in the meanings of (environmental) justice that activists use in other contexts and causes. These divergent notions of justice have become an increasingly prominent form of U.S. social activism, forming what I call a *triple movement*. The term refers to a collection of social change organizations that privilege free market well-being and accept ecological and social risks to assure market freedom. Triple movements mobilize in conditions of persistent poverty and social dislocation and in contexts where free markets have often become part of a given community's collective social fabric. These markets are thus perceived by residents who mobilize sites of acceptance as providing ultimate solutions to community development issues, both economic and noneconomic.

Prominent economic historian Karl Polanyi and his double movement concept have become increasingly relevant under neoliberal hegemony and provide an important foundation for my triple movement concept. Polanyi proposed that free market systems of post–Industrial Revolution capitalism (that is, systems less encumbered by regulations or state interference) would create unprecedented social dislocation and environmental degradation. (Neo)liberal ideology, embodied by laissez-faire capitalism in contemporary economic systems, comprises the first half of Polanyi's double movement; this "first movement" attempts to disembed, or free, markets from regulations and social protections. According to neoliberalism, markets left to function without state regulation provide the most efficient, profitable forms of business and social organization. Social and environmental protections should not hinder growth of market systems. In theory, markets overcome human shortcomings because they behave rationally, respond to clear price signals across cultural and national contexts, and allow unlimited expansion.

Polanyi, however, observed that free markets degrade social and environmental systems when privileged above them—and would lead inevitably to a "double movement" in response. In fact, disembedded free markets are utopian fictions for two reasons. First, political institutions such as the federal state must initially create and regulate systems in which commodity markets thrive. For example, without state laws enforcing private property rights, capitalist market systems as we know them could not operate. Second, people would be so dislocated by free markets' effects on their lives that they would mobilize to limit market expansion and self-regulation. People experience social dislocation in neoliberal, capitalist systems because, while markets in land, labor, and money facilitate free trade, they are not produced by people for trade and are not particularly predictable when traded in market systems. Free markets are thus dehumanizing, unpredictable, harmful to communities and ecosystems, and unresponsive to social needs. As Polanyi explained, "no society could live for any length of time unless it possessed an economy of some sort; but previous to our time no economy has existed that, even in principle, was controlled by markets. . . . Such an institutional pattern could not function unless society was somehow subordinated to its requirements."[96] In neoliberalized market-based economies, temporary, exchange-based social relations dominate, reciprocity and social protections are no longer the norm of exchange relations, and "communities must succumb . . . [as] human society becomes an accessory of the economic process."[97]

Eventually, under extreme stress due to social dislocation (that is, people's experiences of powerlessness or vulnerability related to free market instability), double movements mobilize: activists fight to regulate markets and protect their communities from economic systems that are disembedded from social and environmental contexts. Polanyi predicted that activists would reject commodification of land, labor, and money because marketizing these "goods" was inherently destabilizing. Instead, through the double movement, people would mobilize to demand state-sanctioned regulations protecting society from inherently unstable markets. For example, when energy markets freely orchestrate global trade via abstract price signals, regardless of external considerations such as poverty, pollution, or destabilizing boom-and-bust cycles, they directly affect people's quality of life and community social fabrics. In such cases, Polanyi believed, activists would fight to re-embed markets in their social and environmental contexts, demanding that the federal government create and enforce protections against pollution, economic destabilization, and poverty. While movements could be progressive, regressive, or otherwise divergent in their politics, movements would consistently protect societies from markets that induced instability, unpredictability, and unidimensionality.[98]

Activists in various movements, perhaps especially those mobilizing environmental justice sites of resistance, have seemed to bear out Polanyi's predictions about the double movement.[99] They fight unjust outcomes related to industrial production and its risks and they fight the privileging of free markets above community and ecological well-being. So why have sites of acceptance mobilized in places such as Nucla and Naturita? Why do activists in these communities use alternative notions of environmental justice? I believe that notions of justice change in neoliberal contexts further stressed by key material conditions: persistent poverty, spatial isolation, natural resource dependence, and strong identification with an industry. The modes of social activism they motivate do not resemble sites of resistance or larger double movements. Instead, these sites of acceptance, which often become institutionalized as social movement organizations, become elements of triple movements; here, free markets are trusted as the ultimate arbiters of social equality, and markets are protected as part of the social fabric of the community, despite their globalized volatility. Sites of resistance that contribute to Polanyi's conception of the double movement still mobilize—for example, in fighting structural inequalities that shape environmental injustices. Yet in communities where alternative notions of justice and dominant modes of activism form sites of acceptance that comprise larger triple movements, environmental and community degradation become normalized costs of economic growth.[100]

As I demonstrate throughout this book, many activists have absorbed these alternative notions of justice. They do not demand transformative social change, and their goals for social mobilization have been deradicalized, largely due to people's reduced economic security, shrinking social safety nets, and their identification with industries central to their communities. In this way, social change organizations comprising the triple movement, such as sites of acceptance in Nucla and Naturita, represent the potential power of neoliberal hegemony. Citizens see commodity markets as integral parts of the community and environmental justice as their right to use the wealth within local landscapes to diminish community-wide persistent poverty.

Thoughts on Sociological Relevance

Notions of justice can shift dramatically, especially in communities faced with persistent poverty and spatial isolation. Thus, while such notions may be less transformative, environmental justice researchers must be less rigid about our working concepts of justice: "The question should not be who is the best judge of a concept of justice—activists or theorists? . . . Different discourses of justice and the various experiences and articulations of injustice inform how the concept is used, understood, articulated, and demanded in practice; the engagement with what is articulated on the ground is of crucial value to . . . development of the concepts we study."[101] Identifying patterns in divergent sites of resistance and acceptance clarifies the varied ways in which neoliberal logic can significantly affect social activism and notions of environmental justice across locations and classes. More generally, these interrogations can help us identify and verify the core traits of organizations that form triple movements: namely, faith in corporate and market self-regulation, the privileging of free markets because they are part of individual and community identities, and the assumption that environmental justice means autonomy over local land use to reduce community-wide persistent poverty.

Further, "the justice demanded by . . . environmental justice [activists] is really threefold: equity in the distribution of environmental risk, recognition of the diversity of participants and experiences in affected communities, and participation in the political processes which create and manage environmental policy."[102] Interrogating shifts in notions of environmental justice that accommodate the experiences of persistent poverty and spatial isolation in uranium communities illustrates mechanisms that mobilize sites of acceptance of free market privilege and the risks accompanying industrial production. More broadly, recognizing diversity in notions of environmental justice helps explain how and why activists' goals may shift as neoliberal structures, policies, and logics interact with social fabrics, communities, and ecosystems. If we neglect to explore notions that do not fit tidy definitions of environmental justice, we ignore important ways in which neoliberal hegemony affects communities and environments and sometimes creates paradoxical outcomes for environmental activism. We also miss opportunities for broader sociological explorations of the ways in which outcomes in uranium communities parallel those in communities that depend on other natural resources. For example, in my fieldwork in northeastern Pennsylvania communities with rapid, expansive natural gas development, sites of acceptance have mobilized alongside sites of resistance, supporting rural industrialization and arguing for local control over land use—largely to counter regional persistent poverty, natural resource dependence, and spatial isolation.[103]

The stakes have changed in uranium communities since the last boom busted in the early 1980s. The dark side of neoliberalism's celebration of individual freedom is its power to compel individuals to feel solely responsible for all successes and failures. In sites of acceptance mobilized in uranium communities, grassroots activists privilege markets, even if that means eventual environmental degradation or exposure to boom-and-bust economic cycles. For residents, neoliberalism's "remoralization of the poor" and its cost-benefit analytical approaches make renewed uranium production seem to be the most rational choice. Thus, neoliberal political measures that encourage states to privatize their natural resources, deregulate, and shrink social programs constrain uranium communities and their residents to support an economic activity with questionable legacies and economic volatility, doing structural violence to people's agency and their social justice goals. In this context, people alleviate their most pressing sources of social dislocation before attending to others, often with little federal state assistance. A triage mentality emerges, with immediate economic needs often trumping longer-term concerns such as environmental degradation or transformative notions of environmental justice.

Looking Ahead

During my years of ethnographic fieldwork and mixed-method data collection on the Colorado Plateau, residents throughout the region shared exceptionally rich narratives about the uranium industry. I met people who described their fear of uranium and others who claimed such fear was irrational, as they told me about community members who drank yellowcake (refined uranium ore) mixed with water for breakfast. I spoke with people who had vastly different perspectives on how industry renewal would affect the region socially, economically, and environmentally. The variety of these perspectives influenced me and my views of the industry in ways I could not have predicted. Early in my fieldwork, I worked with activists in Monticello and felt deep sympathy for residents suffering from environmental injustices and health complications. Yet my subsequent work in the deeply impoverished communities of Nucla and Naturita made me question my earlier biases, and eventually I attained a much more nuanced perspective about renewed uranium development. Given my own fluctuating perspectives, I have interpreted the data I share in this book with great care, concern, and appreciation for residents' differing experiences and expressions of social dislocation and their notions of environmental justice.

In the field, I conducted in-depth interviews with more than seventy people involved in uranium production in various capacities. They included members of social movement organizations that focused on the environmental injustices stemming from the industry's regional legacies; case managers and radiation regulators with the Colorado Department of Public Health and Environment; environmental managers and top-tier administrators with Energy Fuels Resources; activists who accepted and resisted uranium renewal; and various leaders in communities across the Colorado Plateau. These interviews gave me sustained interactions with community members and allowed me to cultivate a nuanced understanding of how people viewed uranium development, the state and corporations, individual and community economic security, and regulatory mechanisms. I also learned how people used or rejected neoliberal logic.

I combined data collected from the interviews with findings from a variety of other mixed-method strategies. Between July 2006 and July 2011, I conducted a series of ethnographic participant observation sessions, which gave me a strong sense of how various uranium community residents experienced environmental justice, natural resource dependence, and neoliberal policies. Additionally, I distributed a household survey querying residents about uranium development's legacies and renewal on the Colorado Plateau so that I could compare perceptions across the four communities that were closest to the proposed site of the Piñon Ridge Uranium Mill. Archival analyses of newspaper coverage and regulatory frameworks reinforced my data and findings and allowed me to deepen my knowledge of community histories, their economic instability, and the development of environmental injustices over time. I detail all of these methods in the appendix.

In chapter 2, I describe uranium production's history on the Colorado Plateau and the various communities implicated in the nuclear fuel cycle. I examine uranium production's boom-and-bust cycles and the subsequent social dislocation and economic insecurity that has arisen in uranium communities since the 1950s. The chapter shows how citizens in Monticello, Nucla, and Naturita came to identify themselves with the industry and how Cold War urgency led to large-scale environmental injustice. In chapter 3, I explore environmental injustice and contested illness experiences in Monticello. I focus on the Victims of Mill Tailings Exposure (VMTE) and the many community members who tie local disease to uranium exposure from Monticello Uranium Mill, despite the federal government's denial of a causal link. I analyze how VMTE's activism represents traditional notions of environmental justice in creating a site of resistance, even as activists feel constrained to tentatively support uranium renewal because of its economic benefits.

Chapter 4 focuses on Energy Fuels Resources and the Piñon Ridge Uranium Mill, two key drivers in renewed uranium production. I describe the uranium milling process, mill permitting politics, and implications for meaningful community involvement in land use decision making. Using activists' own accounts, I offer an ethnographic account of activism in sites of acceptance as well as activism mobilizing sites of resistance. In chapter 5, I add life to Polanyi's concept of social dislocation, illustrating how divergent experiences among uranium community residents relate to divergent notions of environmental justice and to modes of activism that are embodied in sites of acceptance and resistance to renewed production. Persistent poverty, natural resource dependence, and spatial isolation structure social dislocation experiences, which vary according to people's class status and to how embedded a specific community is in global market systems.

Chapter 6 highlights an especially provocative outcome of my fieldwork: residents' narratives about uranium industry regulations. Two distinct threads emerged. People who mobilized sites of acceptance asserted that current regulations are better than previous ones, sufficient to protect workers and community members, and adequately enforced by the state and Energy Fuels. Those who mobilized sites of resistance believe that regulations are inadequate, self-monitoring does not ensure regulatory compliance, and the Colorado Department of Public Health and Environment is incapable of thorough enforcement because of shrinking budgets and capacity. Alongside these narratives, I demonstrate how varied perceptions of the state and private uranium producers influence notions of environmental justice and modes of activism, thus affecting activists' alignments with sites of acceptance or resistance.

In chapter 7, I tie together observations about uranium's underaddressed legacies as they intersect with nuclear renaissance. I show how uranium industry renewal exacerbates pervasive social tensions and divisiveness and argue that the support of many residents who mobilize sites of acceptance emerges from structural disadvantages and material deprivations. Even in sites of resistance, activists shape their goals in largely market-based terms. In an era of shrinking state capacity to provide social safety nets such as healthcare and transfer payments, residents in uranium communities see private institutions such as Energy Fuels as not only employers but also the source of safety nets such as healthcare, social networks, and vocational training. I conclude the chapter with discussions about the triple movement, hegemonic neoliberalism's effects on activism, and the ethics of energy development as they relate to environmental justice. I suggest that an ethic of sustainable energy development should involve smaller-scale renewable energy methods empowered by grassroots collaborations and innovations among regional communities. In the chapters that follow, I look forward to illuminating life in America's uranium communities—spaces that have been nearly invisible until now.

2

Booms, Busts, and Bombs

Uranium's Economic and Environmental Justice History in the United States

> The twenty-first century evokes a new chapter of development on the Colorado Plateau. Rural areas are shrinking as they are overtaken by retirees, vacationers, and those seeking "second homes." There is talk of revival of the uranium industry as energy needs proliferate. . . . Yesterday, we were ignorant, or maybe just plain cavalier, about destroying the environment and diminishing our quality of life. May a new conscience ensure that the past mistakes . . . will not play a part in the future.
>
> —Raye Ringholz, *The Uranium Frenzy*

Uranium has mythical power over the human imagination. Formed by a supernova 6.6 billion years ago, it roils at the core of the earth, spurring continental drift. For centuries, uranium had nominal value in Europe and Asia, where it was used as a coloring agent in porcelain and other ceramics.

In North American, the Ute and Navajo (Diné) peoples extracted red-yellow carnotite ore speckled with deposits of uranium, which they pounded to dust and used to decorate their bodies for war and celebrations. In 1789, German chemist Martin Klaproth formally "discovered" the presence of uranium in a rare, dark, oozing substance known as pitchblende and recognized its potential relationship to strange illnesses among silver miners in the Ore and Erz mountains of what is now the Czech Republic. During the mid-nineteenth century, when gold prospectors on the Colorado Plateau inadvertently uncovered large-scale uranium deposits, they never guessed the value of their find. Until the 1930s, uranium was seen as a throwaway element; early miners referred to it as a "weed."[1]

Although uranium's value was overlooked, radium and vanadium (also found in carnotite ore) did spark booms in the early twentieth century. In the early 1900s, after Marie Curie discovered radium's potentially healing properties, pitchblende gained value as she and her husband, Pierre, developed their research program. After visiting the Four Corners region, she began sourcing most of her radium from the Colorado Plateau, which ignited a brief U.S. boom.[2] Scientists hailed radium as a miracle cure for tumors and malignancies. Even *Cosmopolitan* magazine raved over it, comparing radium to "life, energy, immortal warmth."[3] By 1923, however, explorers found less expensive reserves in the Belgian Congo, and American radium markets busted.

A few years later, European scientists discovered that vanadium could improve steel production by adding tensile strength and increasing its flexibility. Quickly the element became a valuable ingredient in weapons and defense-related construction. As World War II approached and other countries intensified their steel production, prospectors on the Colorado Plateau began combing through discarded piles of ore from the radium boom, searching for vanadium to sell to war contractors and steel producers. Uranium's market value—indeed, its commodification—was not established even then. That all changed in 1938, when Otto Hahn and his assistant Fritz Strassman discovered nuclear fission and observed that uranium nuclei split when bombarded. Immediately the element—once discarded in tailings piles near radium mills—became one of the most valuable commodities in the world. Nuclear scientists assured U.S. leaders that developing uranium and harnessing the power of nuclear fission would help establish the nation's international reputation and power as well as end the escalating war.

Uranium's First Boom

The first domestic uranium boom began in the early 1940s as the United States developed atomic weaponry for World War II. Now that the element had become a vital component of national security, its economic value increased dramatically, sparking a uranium rush on the Colorado Plateau.[4] New businesses in the region catered to booming populations, and fledgling uranium communities experienced unprecedented growth. In an oral history collected by the Rimrocker Historical Society, Naturita resident Betty Rutherford recalled the boom's impact on her own and others' businesses: "The Walkers had their jewelry store. Randolphs had a TV store. . . . There were a lot of businesses in Naturita at the time. Things were really booming. It was starting, too, when Mom and I had the café, and . . . I think [our] best years were when we had [the uranium boom]."

Until the end of the first boom in the early 1960s, only the U.S. government could legally purchase uranium; and the nation's increasingly keen competition with the Soviet nuclear program created an exaggerated level of demand. To supply the new market, the federal government encouraged amateur prospectors on the Colorado Plateau to search for uranium and stake claims. For only one dollar per claim, they could become government suppliers. Federal military demands for domestic uranium supplies were so intense, the U.S. government began granting a 10,000 dollar bonus for each significant uranium strike.[5] The uranium-rich plateau provided sustenance for thousands of rural residents. Newly minted explorers, living in tarpaper shacks without utilities, patriotically scoured the desert landscape in hopes of increasing their economic security and changing their class status. As the only legal buyer of uranium ore, the U.S. government wielded enormous power in shaping the national uranium market and the livelihoods of prospectors and their families.

The Anatomy of Monopsony

Embroiled in World War II and searching for a decisive exit strategy, by the mid-1940s, U.S. military leaders felt great anxiety about securing enough uranium. During the first half of the decade, the Manhattan Project produced the atomic bombs that were eventually dropped on the Japanese cities of Hiroshima and Nagasaki. As the Cold War developed in the second half of the decade, the nation frenetically stockpiled nuclear weapons, creating an unprecedented demand for uranium. Yet Manhattan Project scientists had quickly depleted the uranium-rich ore left from the radium era, and the country could not rely on unstable reserves in the Belgian Congo. One certainty emerged from this political and economic angst: the United States needed more uranium immediately.

The federal government decided to act aggressively to create a domestic uranium market. The Colorado Plateau's abundant reserves of carnotite ore made it a logical choice as the U.S. center for uranium extraction and production. In 1946, Congress passed the Atomic Energy Act, which formed the Atomic Energy Commission (AEC) and gave it the power to act as the sole buyer and regulator of uranium ore in the United States. Using a series of market-based incentives, the AEC encouraged private prospecting to fuel the boom in uranium. The catch was that the federal government could be the only buyer and processor of any prospected uranium.[6] Economists refer to this arrangement as a *monopsony*, where a single buyer has many sources from which to purchase a commodity and thus has leverage within a given market.[7] Monopsonies create power dynamics that mirror those of monopolies.[8] In a monopsony, however, the single purchaser of a commodity controls the terms of trade and largely shapes markets for the commodity, dictating prices for goods and often determining how they will be used or redistributed.

Thus, by creating the U.S. uranium market and positioning itself as the sole buyer of uranium ore, the government thoroughly controlled the domestic uranium market and nuclear weapons production from the early 1940s to the late 1950s. Although various companies (such as the Vanadium Corporation of America) bid to manage mills on the Colorado Plateau, the U.S. government initially and sometimes continually oversaw the operations of private contractors. Moreover, it was able to conceal most activities related to uranium extraction, production, and enrichment because they were matters of national security. Well into the Cold War years, the domestic uranium market was veiled in secrecy, a situation that the monopsonic arrangement simplified.

Sparsely populated, sprawling, and rich in carnotite ore and uranium deposits, the Colorado Plateau was an ideal center for extraction. Not only was uranium plentiful, but the region's isolation ensured security and anonymity for the U.S. government's highly classified atomic ambitions. Local populations, who felt they were keys to national security, kept their activities quiet. In their eyes, uranium prospecting allowed them both to express their patriotism and to enhance their wealth.[9] Throughout the nation, citizens believed that uranium would be a powerful U.S. defense in any confrontation with the Soviet Union, and Colorado Plateau residents felt a special pride and patriotism about their direct roles in providing such a vital resource.

The subject began to dominate local culture and neighbors' conversations on the plateau and helped create a new local consciousness. Various towns began to see themselves as "uranium communities." Local newspapers boasted about the boom. Coverage in the weekly *San Juan Record*, which served the Utah region that included Monticello and Moab, was particularly notable and escalated in tandem with the Cold War effort. In 1949, the paper's front page ran only nine stories discussing uranium. By 1951, the number had grown to eighteen front-page stories about uranium, and by 1952 the count had increased to thirty-four. In the mid-1950s, at the peak of the uranium rush, forty editions of the *San Juan Record* carried multiple front-page stories involving uranium. The element had become part of daily life, an aspect of personal identity for people living on the Colorado Plateau.

As the Cold War intensified, uranium became synonymous with political and economic success on the Colorado Plateau. The *San Juan Record*'s increased coverage illustrated the excitement and hope stimulated by the boom. The newspaper highlighted the bonuses paid out to miners and millers, offered tips on prospecting hot spots, published advertisements for new prospecting equipment, and shared tales of county locals who had hit "uranium gold" in their own backyards. In 1949, the paper featured a five-part piece titled "The Mining Bug," which explained how to be a successful prospector, while calling radiation a "boon to medicine."[10] Charlie Steen, arguably the most successful and notorious uranium prospector of the era, became "the media's delight," and articles about his success triggered many others to take up the hunt.[11] His story became a classic American rags-to-riches tale embedded in the adventurous isolation of the rugged West.

The *San Juan Record* paralleled the uranium rush to the "gold rush of yesteryear."[12] Headlines celebrated the "Extensive Uranium Program in Rich San Juan County Empire."[13] They rejoiced when "AEC Hikes Uranium Prices on Low Grade Ores, Ups Bonuses" and were delighted to announce that "Uranium Miners Paid Over One Million Dollars in Bonuses."[14] Beginning in 1951, editors placed weekly updates about uranium stocks prominently on the second page. In a February 1955 issue, they included an eyecatching 122-page insert about energy development on the Colorado Plateau, touting the virtues of uranium and equating it with progress, the American Dream, and economic security.[15]

As newspaper coverage suggests, uranium became an unparalleled political, economic, and cultural force in Colorado Plateau communities. The oral histories of community members and uranium workers reveal both their economic dependence and their strong cultural identification with the industry. Some of the men who worked in mines and mills during the 1940s and 1950s recalled their pride in the industry and employees' strong work ethic. Calvin, a former miner from Naturita, recalled, "When we drilled ventilation holes for the mines, I could see satisfaction on [my father's] face when he could shine light, from a mirror, down a 400-foot hole and see the bottom. . . . Regardless of whether or not you liked your job, what is important is that when you went home at night, you [could] feel good about yourself, knowing that you plowed a straight furrow or you [could] see to the bottom of the hole."[16] Workers often mentioned their patriotic pride in U.S. technology and the uranium market they had helped to fuel. Naturita resident William Kyle, a longtime prospector and miner, recalled the glow he felt about the small part his family played in market growth: "The uranium boom was just beginning. Nobody knew it yet, though. The government was getting ready to procure some uranium to build the bomb. . . . They needed uranium. . . . So we moved over here and Dad went to work at Uravan for 55 cents an hour, and that was a big raise to what he had been making. . . . We had prospected quite a bit, . . . found a little bit here and there. . . . [One time] Dad reached down and picked up a little piece of rock about an inch thick and he handed it to me, saying 'You know, kid, that looks like ore.'"[17]

An atomic subculture developed throughout the region. Monticello, Moab, Nucla, Naturita, Uravan, and other communities dotting the Colorado Plateau incorporated uranium into their local landmarks and residential centers. Builders of a new subdivision in Moab christened it "Uranium Village."[18] Towns named new roads after uranium and nuclear technology. During the 1950s, young women were crowned as Uranium Queen and Miss Atomic Energy. Moab changed the name of its annual rodeo from Red Rock Roundup to Uranium Days Rodeo. Moab, Grants, and Monticello each had a restaurant called the Uranium Café; and retailers of all kinds began selling Geiger counters and prospecting equipment.[19]

As Moab and its environs assumed the identity of "uranium capital of the world," speculators were establishing a uranium stock market in Salt Lake City. By the early 1950s, they had transformed Utah's largest city into the "Wall Street of the Uranium Market."[20] A largely unmonitored stock market convened here, one that offered penny stocks to the public to fund the growing number of prospecting businesses that were seeking ore on the plateau. This market encouraged experienced entrepreneurs and sparked an influx of investment money that facilitated increased stock and land-right speculations. Shares to companies such as Uranium Oil and Trading, one of many new prospecting companies, were sold "literally over-the counter" to investors lured by the prospect of wealth; the first shares were sold in the back of Whitney's Coffee Shop in downtown Salt Lake City.[21] Few knew much about uranium development, the companies they invested in, or the legitimacy of either; they were driven by excitement over economic success and involvement in new technologies.

After these successful initial offerings, the number of speculation stockbrokerage firms skyrocketed. According to uranium historian Raye Ringholz, "where there had been 20 traders in 1953, by the next spring there were 112."[22] The stock market reached an intensity and ubiquity that few initial investors had predicted; some uranium stocks were even traded on the New York Stock Exchange. Though the bubble would burst along with the boom, for a few years the uranium penny stock market was fueled by a spirit of discovery and the energy of young stockbrokers, then some of the youngest in the profession.

Colorado Plateau residents easily linked uranium development to patriotism, fear of the Soviets, and support for the war effort. Local newspapers reinforced this regional pride—for instance, by publishing articles about the Colorado Plateau uranium that had been used to make the bombs dropped on Hiroshima and Nagasaki. In the *San Juan Record*, speculation about a Soviet attack on the United States mirrored residents' belief that local support for the industry was central to protecting U.S. security: "The attack most surely would be by atomic bomb. Russia would not attack the continental United States without it. At present, the bomb is believed to be a US monopoly, but probably won't stay this way."[23] Widely circulated instructions told people how to survive a Soviet-instigated nuclear war, reminding uranium community residents of their vulnerability and implying that their best recourse was to supply the military with uranium.[24] Articles about a Civil Defense Program in San Juan County reinforced worries that the area could be a Soviet target.[25] Every week the AEC sent out announcements about setting aside public lands for uranium prospecting or developing new methods of uranium refinement "to fight the Axis foes" or the "tyranny of communism."[26] Residents were constantly reminded of uranium's connection to national security and of their own immersion in the market.

Regional journalists recognized that without the monopsony conditions arising from military-industrial demands, uranium would be nothing more than dirt under people's feet. Discussing "Uranium's Lure," an editorial in the *San Juan Record* declared, "'Tis military need that creates the present demand for this mineral that in the gold and silver days was ignored as of little or no value." This prescient editor worried that the federal government's singular purchase of uranium ore for confidential defense projects could mean "years of stagnation to many communities today enjoying prosperity due to military demands for uranium."[27] Most of the communities embedded in uranium markets had a history of involvement in natural resource economies. Thus, with a bust looming in this unstable, unpredictable, government-led arms and energy economy, some residents were feeling the all too familiar pangs of social dislocation, this time connected with uranium production.

The Political Economy of Uranium Monopsony

From the beginning, the uranium market induced persistent forms of social dislocation. Given the unique political-economic context surrounding its production, the U.S. uranium market was disembedded or disconnected from uranium communities' social fabrics and related environmental outcomes as soon as it emerged. Meanwhile, the stock market fueled investment, speculation, and fixation on a uranium price point, even though a culture of secrecy shrouded most market activity. No one formally considered the first boom's effects on residents' quality of life, economic stability, or their environment's integrity. Global conditions drove the market, which was generally self-regulating. There were few checks on worker safety and economic or environmental sustainability; the market's well-being was privileged above people. Though locals prospected for uranium in the land they called home, and though their economies increasingly depended on it, they had little control over the market, which was connected to a globalized set of uranium markets in Europe and Africa.

Given the combined influence of the uranium monopsony and the penny stock market, prospectors, mine and mill workers, and even communities had to deal with economic volatility and broader social dislocation experiences such as employment instability, pressures on infrastructure and social cohesion, and environmental degradation. The AEC was far more interested in price points than in the market's effects on communities such as Monticello, Nucla, or Naturita. Even uranium penny stocks depended on the AEC's price points. Stock values were in jeopardy if prospectors and miners in publicly traded uranium exploration outfits could not receive adequate prices for their mined uranium or if the market changed unpredictably. In the mid-1950s, when the Securities and Exchange Commission finally began examining the penny stock trade, it revealed the precarious financing behind most uranium firms, a situation that made investors even more vulnerable to market fluctuations.[28]

By the mid-1950s, prospectors had helped the AEC garner a surplus of uranium; and by the end of the decade the commission declared that "it [was] no longer in the interest of the Government to expand production of uranium concentrate."[29] The United States had accumulated reserves of 71 million tons of uranium, after producing slightly less than 9 million tons of ore per year. The market was saturated. With the monopsony changing shape, the AEC now announced that it would introduce a modified buying program, purchasing small quantities of uranium ore only from reserves developed before 1958. By the early 1960s, the market had gone bust. On the Colorado Plateau, employment opportunities decreased and destabilized. Populations shrank, medical clinics and restaurants locked their doors, and mills and other uranium-processing facilities shut down.

Among those facilities was the Monticello Mill, which the AEC had owned throughout its production years. When the commission announced the mill's closure in 1959, community members formally organized a protest. As I will discuss, even then residents were concerned about the industry's environmental health effects. But losing the mill and the economic security it brought to the town meant losing an integral part of local identity, and the era's sustained community support for the industry foreshadows the toxic ambivalence about uranium production that still exists in most uranium communities.[30] For the first time, the AEC came under fire for operating "under a guise of 'secrecy' . . . [and] establishing itself as an all-powerful bureaucracy whereby it can dictate, without rhyme or reason, as it sees fit."[31] Mill workers, along with prospectors and independent miners, began to characterize the commission as a "dictatorship in a democracy."[32]

Shortly after the AEC announced the closure of mills in Monticello, Naturita, and elsewhere, the *San Juan Record* began lamenting the loss of local jobs. One editorial pointedly bemoaned the dislocation that uranium community residents were feeling:

> Free enterprise in the uranium industry has faced the AEC firing squad. The industry is at their mercy. Miners are told how much ore they can mine; where and if they can sell it; and for how much. Dictates to the mills tells them whose ore they can process; how much; and for what price. The AEC holds the rifles, and it's too late to bargain with the executioner. . . . As is true with most government bureaus, the AEC is determined to prolong its own life—save their own jobs. Future years in the Atomic Age will find America's development and progress strangled. AEC has already killed concern for American uranium mining and milling. Continued huge purchases from Canada and overseas while at the same time discouraging and fostering cutbacks and discontinuance of American production, leaves US uranium men in a helpless, hopeless position.[33]

In short, while the U.S. government and economy benefited immensely from commodifying uranium and using it to rapidly develop military and energy technologies, communities incurred most of the industry's long-term costs. Although the first uranium boom brought temporary wealth, excitement, and patriotic fervor to the Colorado Plateau, its bust led residents into economic desperation, insecurity, and powerlessness, reminding them of their spatial and political isolation and leading them to distrust the federal government. Their economic instability and related social dislocation hinted at future dislocations, ones that were exacerbated by health complications and environmental injustices.

Uranium's Second Boom and Permabust

When the AEC amended the Atomic Energy Act in 1954, it distanced itself from an industry it had created.[34] The amendment gave private corporations incentivized access to uranium reserves and markets, offering various subsidies and protections as these utilities acquired the materials and technology necessary to generate nuclear power. Private companies entering the nuclear industry spent abundant time and money on strengthening public perceptions of nuclear power as a clean form of energy, a means to "promote world peace, improve the general welfare, increase the standard of living, and strengthen free competition in private enterprise."[35] Nonetheless, as a second uranium boom began in the early 1970s, the federal government continued to play a strong role, functioning now as the main regulatory agency overseeing uranium production. Private energy companies had become the leaders on the ground.

By the mid-1970s, driven by commercial interests in nuclear power, increasingly consolidated energy companies had stimulated demand for uranium.[36] Communities on the Colorado Plateau welcomed the new boom; but even with legislative and financial support, it was short-lived. Despite the industry's new clean-energy sheen, environmental outcomes were the most significant barriers to its success. Although extraction and production themselves created problems, accidents at nuclear reactors were the primary reason for the industry's second bust. In 1979, the infamous near-meltdown at Pennsylvania's Three Mile Island reactor spread fear throughout the United States, which was reinforced by the Chernobyl disaster in Ukraine.[37] No new U.S. reactors were sited after the Three Mile Island incident, yet cheap sources of uranium were still available from African, European, Canadian, and other global markets. So by the late 1970s, demand for U.S. uranium again fell.

In spite of its brevity, this second domestic uranium boom signaled the growing dominance of private enterprise and a renewed focus on energy generation as opposed to weapons production. Community well-being and economic stability were not key concerns as corporations entered the U.S. uranium trade. Because these private entities were responsible to shareholders and global commodity chains, they were driven by profit margins and technological innovation in energy development. With uranium markets no longer encumbered by government-set price points, market stability was totally at the mercy of global dynamics and the "invisible hand." Once again, communities such as Naturita, Nucla, and Monticello were vulnerable to the whims of self-regulating markets, this time in an increasingly globalized context. While community members experienced social dislocation after the second bust that paralleled their experiences during the first, the second downturn was much lengthier, and the effects were even more pernicious. Between the early 1980s and the mid-2000s, populations dwindled, community infrastructure decayed, and local identification with uranium was tainted by environmental degradation and injustice.

Uranium Communities, Natural Resource Dependence, and Social Dislocation

Residents of uranium communities felt each bust at their kitchen tables and in their neighborhoods' diminishing quality of life. Thanks to the industry's rapid rises and descents, residents were forced to deal with widespread employment instability and persistent poverty. Powerless against extralocal market fluctuations, people also experienced broader cultural dislocation: uranium no longer enhanced their social status or provided economic security.

Naturita and Nucla, Colorado

Naturita and Nucla are separated by about four miles; since their founding, they have depended on natural resource extraction for economic stability. The Colorado Cooperative Company founded Nucla in the late nineteenth century with the goal of creating a communal, rural atmosphere away from the chaos of Denver. Naturita, founded in 1881, expanded around ranching, agriculture, and extractive mining activities. Located in arid, high-elevation, southwestern Colorado, the communities worked together for years to build an extensive irrigation ditch system, which has fortified their sister-city relationship.

In the early years, economic security meant subsistence agriculture, and a few pioneer families settled in the area to grow fruits, vegetables, alfalfa, and eventually raise cattle on sprawling ranches. But geographical remoteness kept residents economically isolated, prohibiting any extensive involvement in national agricultural markets. All of this changed in 1898, when prospectors found carnotite ore while exploring for gold. Both communities, but particularly Naturita, became embedded in national and international markets for radium, vanadium, and then uranium. As their natural resource dependence developed, the towns' economies became more oriented to global market dynamics. In 1914, to facilitate new mining, Standard Chemical built a five-mile road connecting Naturita to coke ovens that were processing coal. By 1930, a company called Sterns and Rogers had constructed the Naturita Mill for Rare Minerals, which, on an irregular basis, processed regional vanadium ore that was then distributed to the eastern United States for steel production. By 1939, however, the Vanadium Corporation of America had refurbished the mill and regularly produced and exported vanadium to meet the increased global steel demands of World War II.

By the late 1930s, the Naturita Mill, now the area's largest employer, had also begun producing uranium.[38] In the late 1930s and early 1940s, most of this locally processed ore went to the Manhattan Project. After the project ended, the mill closed for a year but reopened in 1947, thanks to a contract between the Vanadium Corporation of America and the AEC, which was accumulating uranium for its Cold War nuclear weapons arsenal. By 1955, the mill was producing more than two hundred tons of uranium ore each day, and Naturita and its neighbor Nucla had become uranium boom towns. Between 1939 and 1945, Naturita's population grew from fewer than 100 residents to more than 1,200 people.[39] Along with this burgeoning population came numerous new businesses in Naturita and Nucla, including a gas station, a food market, a hotel, a bar, a restaurant, a drilling company, a theater, and trailer courts for transient mill employees.

But change was on the horizon. When the AEC's demand for uranium diminished in the late 1950s, the Naturita Mill closed. Although the commission did reopen the mill in 1960 as an experimental uranium and vanadium concentrating facility, it closed the facility permanently in 1963. Yet even during boom periods, and even when the AEC was the sole purchaser of uranium ore, residents were aware of the tenuous quality of their lives. Pat Daniels, a lifelong Naturita resident and former uranium miller, recalled, "Down at Grants, New Mexico, they could produce uranium for a lot less money than they could up here. They had huge deposits. . . . [Naturita and Nucla] just couldn't compete in the other markets that had sprung up. The price was the big thing."[40] Local historian and lifelong Paradox Valley resident Marie Templeton noted that many people have accepted boom-and-bust cycles as a normal part of life: "Uranium has been a part of our economy since the early 1900s, good or bad. It has either been a boom phase or a bust phase. Seems like there is no middle ground."[41] Still, when the boom times were in full swing, the lure of the market was hypnotic—and rewarding. According to former uranium miner Bob Barker, "we were living on Snob Hill at the time [of the first boom]. Two dollars and twenty-two cents an hour was the wage. . . . I worked six days a week, paid all the bills, and had money left over. Everyone was working. A $3,500 AEC Bonus Plan was in effect. Every mine in the area was producing."[42]

As uranium markets destabilized, Nucla and Naturita largely returned to their agricultural and mining-supply roots. Yet they retained the infrastructure and workforce to extract and process ore when the market demanded it. After the brief boom of the 1970s, the communities spent much of the next two decades mired in local economic recession, with most of their remaining jobs linked to coal mining, cattle ranching, and power production at the local coal-fired plant. By 2011, Naturita's population had dwindled to 545.[43] Although the median household income in Colorado was roughly 55,000 dollars in 2009, Naturita's was only 22,430.[44] Nucla, while mired in the same boom-bust cycle, had a stronger link to service jobs in Telluride's tourism industry. By 2011 its population had dropped to slightly more than seven hundred people, but service-sector employment kept 2009 median household income at a more typical 48,268 dollars.[45]

Uravan, Colorado

Before its incorporation as a town, Uravan was the site of the Joe Junior Mill, a high-grade radium, vanadium, and uranium mill built in 1912 in southwestern Colorado's Montrose County. The first important customer was Marie Curie's research team, which was headquartered close by. Standard Chemical owned the mill until 1929, when the U.S. Vanadium Corporation (bought by Union Carbide in 1926) purchased both the land and the facility. The original mill was torn down, and U.S. Vanadium constructed the new Uravan Mill in 1936.[46]

Uravan was founded in 1936 as Union Carbide's company town, and the community's instant population influx accentuated the region's social and economic energy. During the boom of the 1940s and 1950s, Uravan's tree-lined, idyllic neighborhoods housed slightly more than eight hundred people. The community had multiple schools, a small hospital and medical facilities, and swimming pools and parks—all financed by Union Carbide.[47] Uravan was the company's little jewel in Paradox Valley, culturally and patriotically bound to the uranium industry, which Union Carbide ads framed as holding the "promise of a golden future . . . [and] thriving new communities."[48]

According to rumors, the Uravan Mill provided some of the yellowcake (refined uranium ore with a dustlike consistency) used to construct Fat Man and Little Boy, the Manhattan Project atomic bombs dropped on

Hiroshima and Nagasaki.[49] During the Cold War, the mill continued to play a pivotal role in uranium production. Between 1936 and its final closure in the mid-1980s, the Uravan Mill produced more than 40 million tons of uranium and vanadium.[50] In the meantime, both the mill and the community were weathering the instability of uranium's booms and busts. Between 1981 and 1984, when the mill was operating only 60 percent of the time, Uravan's population dwindled and the once-bustling downtown faded. During these bust years, Dow Chemical's subsidiary, Umetco, took possession of the mill—and in the process acquired a legacy of severe radioactive contamination. In June 1986, the facility and the town were declared a Superfund site. The cleanup required drastic measures to protect surface and groundwater and minimize radon emissions.[51] Workers evacuated all town residents, remediated contaminated soil and material throughout the four-hundred-acre site, relocated more than 3 million cubic yards of mill wastes and contaminated materials, constructed repositories, collected 70 million gallons of seepage from Uravan's hillside and near tailings, and extracted more than 245 million gallons of contaminated liquid from groundwater.[52]

By the time remediation was complete in 2001, Uravan was destroyed: workers contracted by the federal Superfund program demolished and burnt 50 major mill buildings and all of the town's 260 structures. Today, only carefully placed stones show where a company and a community used to thrive. Many former residents now live in Naturita or Nucla, but they still ache for their home.[53] Marie Templeton, who lived in Uravan between 1949 and 1952, warmly recalled the community and the amenities that Union Carbide provided: "We had a ball team, and the Red Cross sponsored swimming lessons for the whole area. All summer long, everyone would get together and have weenie roasts and marshmallow roasts."[54] For the many residents who still identify with uranium production, Uravan's demolition was a cultural catastrophe. Even though public health complications related to long-term uranium exposure such as cancer clusters and respiratory and lung ailments have emerged in the region, strong local identification with uranium has meant that large numbers of residents continue to deny that there is a link between cancers and uranium exposure. As I mentioned in chapter 1, a few local characters even claim that a daily dose of yellowcake mixed into a glass of water keeps them healthy. Having already lost so much, they are unwilling to accept any other negative assessments of the industry.

Monticello, Utah

Monticello is located in southeastern Utah, about sixty miles south of Moab and just east of the state's sprawling red rock landscape. In the late 1930s, government officials selected it as a site for one of a handful of U.S. uranium and vanadium mills. As I will discuss in later chapters, the Monticello Mill refined uranium through various leaching processes.[55] Though it initially processed vanadium for use in wartime steel production, it began producing uranium-vanadium sludge for the Manhattan Project in 1943 and continued to do so during the Cold War.[56] The mill was unique because it was always owned by the U.S. government; in contrast, the ten other U.S. mills that operated concurrently were owned by private industry for at least part of their existence.

The mill brought much-needed employment to the community, but by 1960 the price for uranium ore had fallen so drastically that the government decided to close the facility. Monticello was left in the dust, literally and figuratively. Mill operations had left severe contamination, enough to create two designated Superfund sites.[57] In the meantime, residents began to notice health abnormalities throughout the community.[58] According to the federal Agency for Toxic Substances and Disease Registry, the cancer rates in San Juan County, and in Monticello in particular, exhibited unusual patterns.[59] However, because of the town's small population, epidemiological tests of significance for elevated rates of disease did not substantiate residents' experiences. The agency thus concluded that uranium exposure had not led to increased rates of cancer and other diseases in the community.[60] As I will discuss in chapter 3, the organization known as Victims of Mill Tailings Exposure, which mobilized in Monticello in 1993, believes otherwise.

Moab, Grants, and Chapters of the Navajo Nation

Moab in Utah, Grants in New Mexico, and several Navajo Nation chapters (communities) in the Four Corners region also helped to establish the uranium industry on the Colorado Plateau. During the first boom, Moab became a mining and milling center known for its rugged prospectors and conservative Mormon population.[61] Charlie Steen moved his family to Moab in 1953 after claiming Mi Vida Mine, the biggest uranium strike on the plateau. His good fortune catapulted him to national fame. In the public imagination, "uraniumaires" were scrappy underdogs achieving the American Dream.[62] As Steen remarked, "poverty and I have been friends for a long time, but I'd just as soon keep other company."[63]

With his success drawing attention to Moab, the agricultural village rapidly transformed into an industrialized, bustling uranium boomtown. Between 1950 and 1956 the population increased from 1,200 to more than 4,000, and town officials lamented the lack of housing and other basic requirements.[64] Tent cities and trailer parks soon inundated Moab's riversides and outskirts, and locals even rented space on their front lawns to prospectors.[65] New subdivisions and hotels were constructed (many by Steen), but the booming population overburdened schools, sanitation systems, roads, public utilities, and retailers. Traffic on Main Street increased by 300 percent between 1953 and 1955.[66]

Like other uranium communities, Moab began to develop and embrace its atomic subculture.[67] It named a new subdivision "Uranium Village," retailers opened uranium jewelry stores on Main Street, and a sign welcomed visitors to "The Uranium Center of the World."[68] In 1957, when Steen opened the Uranium Reduction Company Mill, the boom was at its peak in Moab. By 1961, however, Steen had sold his shares in both the mill and his mine to the Atlas Corporation for 25 million dollars, signaling a downturn in the market.[69] Concerned about a bust, town leaders and residents began to diversify their economy, focusing on potash mining and especially tourism.[70] Given Moab's proximity to two national parks, public lands, and abundant recreation opportunities, this transition to tourism worked well, helping to assuage the effects of future busts. More recently, the town has constructed extensive, high-quality mountain bike trails, which has added a liberal, scrappy, western vibe to otherwise conservative southern Utah. Today Moab is world renowned as a mountain bike mecca and is a gateway to Arches and Canyonlands national parks. Still, uranium's environmental and economic legacies persist, even if they are minimized to avoid tarnishing the town's tourist reputation.[71]

In New Mexico, the town of Grants also weathered boom-bust conditions. In 1950, a Navajo sheepherder named Paddy Martinez found a substantial uranium deposit near Haystack Mountain, which drew in "swarms of prospectors."[72] The result was multiple mines, ore-processing mills, and AEC ore-buying stations. Although for a time Grants rivaled Moab as the uranium capital of the world, it, too, suffered from dependence on unstable markets; the initial boom also led to overburdened housing, schools, utilities, and retailers. As in other communities, uranium shaped local culture; Grants even hosted a Uranium Queen pageant in which the lucky winner received ten tons of uranium ore as her prize.

The Navajo Nation, which spreads across a vast portion of the Four Corners region, holds massive deposits of uranium ore, and many of its chapters have contended with boom-bust pressures.[73] Today, about 270,000 people live on about 26,000 square miles of reservation, where poverty is deeper and more persistent than it is in other uranium communities: 43 percent of Navajo (Diné) people currently live under the poverty line, and 42 percent are unemployed.[74] Although 51 percent of the population still relies on mining as a main source of income and employment, the reservation's natural resource dependence was even more pronounced during the boom of the 1940s and 1950s. In those days, Diné chapters primarily supplied a reserve labor force for uranium mines and mills; wealth was concentrated in predominantly white communities such as Moab.[75] Dangerous working conditions were the norm for Diné miners and millers, whose labor was less regulated than that of white workers in other pockets of the Colorado Plateau.[76]

Radioactive contamination continues to plague Diné people and Nation chapters. On July 16, 1979, the United Nuclear Corporation in Church Rock, New Mexico, precipitated a massive tailings pond spill that released the largest volume of radioactive waste in U.S. history. The spill remains the second most radioactive release in human history.[77] Today, more than 1,200 abandoned uranium mines on Navajo land remain unremediated, and as many as 500 of them may require cleanups that will cost tens, even hundreds, of millions of dollars. In response to these lingering legacies, the Navajo Tribal Council passed the Diné Natural Resources Protection Act of 2005, which issued a moratorium on all uranium mining and milling activity on the reservation.[78] Though people are divided about this legislative measure, supporters contend it will end uranium's boom-bust cycle and its detrimental economic, environmental, and social effects on Navajos.

Environmental Justice Legacies in Uranium Communities

As the federal government shaped uranium markets and changed the social fabrics of uranium communities, it also left environmental legacies.

Degradation was extensive and included radioactive contamination of soil and groundwater, widespread use of tailings waste to build community infrastructure, and residential uranium exposure. At first, community members had little knowledge of or control over such contamination; but as environmental injustices and health complications became visible, residents have mobilized to reveal and address them. Nonetheless, problems continue to linger: cancer clusters, reproductive struggles, lung diseases, and other ailments, all of which have contested connections to uranium production.

Early Knowledge of Uranium's Occupational Dangers

Long before the uranium rush on the Colorado Plateau, Europeans had had extensive experience with the element. Reports from as early as the 1500s recorded strange deaths in regions of Germany and Czechoslovakia where silver, bismuth, nickel, and other materials were mined—the same areas from which the Curies later sourced their pitchblende. For centuries, miners in the region had developed a fast-moving illness known as *Bergkrankeit*, or the "mountain disease."[79] Epidemiological studies in the early twentieth century (still some of the most advanced and best-documented studies of occupational illness) concluded that deaths of hundreds of miners in Germany's Erz Mountain region could be connected to exposure to pitchblende and the radioactive uranium it contained.

As early as 1942, American health officials familiar with the European findings warned that uranium mining and milling might have negative environmental and health outcomes. Ralph Batie, chief of health and safety for the Colorado Raw Materials Division of the AEC, had scrutinized the Erz Mountains epidemiological studies and was concerned that the miners under his watch would have similar experiences if officials did not shield them from excessive exposure to radiation. Batie shared the European findings with the U.S. government, the Vanadium Corporation of America, the U.S. Vanadium Corporation, and the Climax Uranium Company, recommending that all install better ventilation equipment in their facilities. But the federal government's focus was on the war and national security, and his suggestions were rebuffed.[80]

In 1949, Batie invited Duncan Holaday, a radiation expert with the U.S. Public Health Service, to his Grand Junction office to review the European studies and preliminary findings about fatal kidney inflammation among some Colorado Plateau miners.[81] According to Batie's later deposition testimony, he told Holaday that "at some of the more dusty operations in the mills, the concentration of air-borne alpha emitters [radioactively charged particles] was several thousand times as high as those permitted in other AEC installations. . . . Readings indicate the probability that severe internal radiation hazards existed in many operations."[82]

Shocked, Holaday insisted that the matter required intense scrutiny, and he eventually moved to the plateau to conduct extensive research and advocate for worker safety and health. He conducted his own field tests, often entering uranium mines and mills to take air-quality measurements and observe occupational conditions. What he found disturbed him: mines and mills with few ventilation shafts, workers without respiratory equipment or proper protective clothing, and radiation concentrations often thousands of times higher than acceptable levels.[83] Diné community leader Luke Yazzie told Holaday that conditions in Navajo Nation mines and mills were even worse: workers received no preliminary medical exams, neither showers nor clean clothes were provided, and the scarcity of drinking water meant that Navajo men had to drink mine and mill water contaminated with uranium.

In his 1952 report for the U.S. Public Health Service, Holaday reviewed the European studies, reported findings from his own fieldwork, and recommended that the federal government take the lead in worker protection, specifically by digging more ventilation shafts, installing fans and ventilation equipment, and communicating with miners and millers about potential health effects and the importance of respiratory safety. He strongly recommended that mines and mills provide respiratory equipment and require workers to use it. As he discovered, however, the federal government did not share his concerns about worker safety and exposure levels. Instead, national security remained the top priority. Despite Holaday's urgent pleas, conditions remained much the same, even into the 1960s.

In the meantime, workers remained ignorant of the dangers. Why didn't Holaday share his findings with them? After the release of his 1952 report, the AEC had made him promise not to tell anyone in the uranium communities about his discoveries; otherwise, he would lose access to the mines and mills he was studying. He could continue his fieldwork only if he told workers they were simply being monitored to ensure continued good health, a service generously provided by the U.S. government. For years, Holaday struggled with this ethical dilemma. Should he continue to conduct his classified health studies and record disease development among uranium workers, or should he disclose his intentions and findings and risk losing his access? Constrained by the government's power, he resigned himself to keeping his findings confidential as he continued to study health outcomes. Consequently, he was unable to alert workers about issues such as lung cancer and noncancerous pulmonary diseases, which he suspected were linked to uranium and radon exposure.[84]

Holaday's safety concerns were justified. Only a few years later, an AEC Health and Safety Laboratory study of 215 workers at the Monticello Ore Concentrating Plant revealed that facility had no effective dust-control measures. Eighty-six employees had been exposed to dust concentrations above the maximum allowed—in some cases, five times higher.[85] Many uranium workers were taking yellowcake home with them on their work clothes, which were then often washed with the family clothes. In later years, a number of workers did develop lung cancer and other ailments, and some were compensated.[86] Yet the government's unwillingness to install ventilation systems or to allow workers access to Holaday's research had already demonstrated that the uranium market was more important than human health.

Early Social Dislocation and Environmental Justice Mobilization

Uranium mills, which were often located near communities, created abundant dust, smog, and other pollutants, which residents perceived as vaguely threatening, despite their pride in the industry. As early as 1945, activists mobilized sites of resistance to the risks of uranium production. In San Juan County, for instance, residents complained of strange fumes from the roaster stacks of various mills, which seemed to correspond with paint peeling on cars and holes eaten into screen doors and clothes hung outside to dry, as well as with breathing problems and unexplained livestock deaths.[87] In April of that year, a superintendent working for the Vanadium Corporation of America received a petition "signed by many townsfolk" and "asking for a check on the deadly sulfuric acid fumes which again spread through [Monticello] now that the mill is in operation."[88]

Responses to residents' environmental health concerns mirrored those that Holaday had received. As the Cold War escalated, the federal government dismissed and minimized anxieties about dust levels and air quality. A guest editorial in the *San Juan Record* asserted that "all that dust people complain about coming from the mill" was being exaggerated and might actually be a positive outcome because "dusting takes up our wives' time . . . [and] keeps [them] in better physical condition."[89] The dust also kept nearly every industry in town in business. A decade later, concerns about dust levels and effluents from mill roaster stacks were still being portrayed as illegitimate, irrational, and insignificant.

While the mills operated, their production activities remained classified, as did contaminants created at the site or released into the air. When the mills closed and sites were abandoned, they were often enclosed in barbed wire. But such fences were poor deterrents, especially for children tempted to play on a tailings pile or explore a mill site. At various times between the late 1990s and 2007, the Victims of Mill Tailings Exposure distributed questionnaires to community members in Monticello, Utah. People's responses revealed patterns of social dislocation caused by both environmental degradation at these sites and the lack of government transparency about contaminants. Residents' comments vividly portray a sense of powerlessness and vulnerability. One person who lived in mill housing when the facility was operational recalled the ore dust that frequently flew through her home and noted that her "family . . . suffered many diseases, various cancers, lung diseases, chronic headaches, and chronic coughs the past 35 years." Another reported, "I got breast cancer [early]—no one in my family ever had it, the doctors were surprised at my age and not having any problems in my family." Community members spoke of family members who had died of cancer early in life; one resident disclosed that "this we believed was caused by mill fumes." A woman said "that my husband and son would still be alive if not for the uranium mill." Two surveyed parents reported that their children frequently played on the tailings piles and swam in the nearby pond, which the parents connected to frequent "colds and nose bleeds." These comments are only a few of many similar responses.

The first case of documented illness originated in Uravan in September 1956, when Tom van Arsdale, a Nucla resident and uranium miner, was diagnosed with lung cancer. Victor Archer, who had replaced Holaday as director of the uranium workers study, recalled that even then he had a feeling that van Arsdale's case was the first of many that would surface.[90] He was correct. Van Arsdale's illness and eventual death were caused by abnormal oat cells, which had created a small-cell lung cancer similar to that observed in uranium miners in Germany and Czechoslovakia. In 1957, Robert Johnson, a uranium miner in an inadequately ventilated facility, was also diagnosed with small-cell lung cancer. In that same year, a National Cancer Institute study made the startling discovery that 65 percent of tested miners had cancer.[91] This study, however, remained confidential for reasons of national security.

As these initial health concerns surfaced, the uranium industry navigated its first bust. To save money, the AEC decided to reduce its budget for regulatory enforcement and cut regular inspections to thirty state-owned uranium facilities, despite Holaday's protests that these cuts would damage the already questionable quality and quantity of regulatory oversight.[92] In the meantime, more and more cancer cases were surfacing among uranium workers; and revelations were emerging about atomic bomb tests at the Nevada Nuclear Test Site and widespread exposures to radioactive fallout throughout southern Utah, northern Arizona, and parts of Nevada.[93] Residents' feelings of social dislocation and distrust in the U.S. government were exacerbated by their discovery that the Department of Defense had scheduled bomb tests to coincide with wind patterns that would direct fallout over these sparsely populated regions—what documents referred to as areas inhabited by "low-use segments of the population."[94]

Cancer Clusters and Environmental Health

In the 1960s, as uranium's second boom began, both workers and community residents began noticing unusual health conditions in their families and neighborhoods, including elevated rates of certain cancers, respiratory and other lung complications, and general health abnormalities. Various health studies corroborated these perceptions.[95] Monticello had an unusually high number of leukemia cases among elementary- and secondary-age children.[96] Many additional victims were adults even if exposed as children because uranium-related illnesses typically have a latency period of about thirty years.[97]

Wives began to note patterns in their husbands' cancers and deaths as well as in their own health abnormalities. They linked these outcomes to their husbands' extended uranium exposure in mines and mills and to the dust they brought home on their work clothes and gear. Eola Garner was one of the first widows to fight for compensation after her husband, Tex, a uranium miner, succumbed to lung cancer in 1963.[98] Tex had worked at the Uravan, Naturita, Dove Creek, and other uranium mills. Before his death, he told Eola that he believed his cancer had come from working with uranium. An autopsy confirmed that his bones had elevated levels of lead-210 and radioactive polonium, indicating prolonged and excessive exposure to radium and radon daughters. Calculations confirmed that Tex had been exposed to 1,870 working-level months of radiation, well above safe exposure levels.[99] Eola filed lawsuits in Utah and Colorado, asking for compensation for her husband's death from long-term exposure to uranium. Though the case dragged on for years and earned her little in the end, her activism motivated other widows, primarily in Utah and Arizona, to work for environmental justice and mobilize sites of resistance.

Similar environmental injustices and health problems occurred on Navajo land. Before the 1950s, there were no recorded cases of lung cancer in Navajo chapters, but this changed after the uranium booms.[100] By 1969, uranium miners were twenty-eight times more likely than other Navajo men to develop cancer. Between 1969 and 1993, two-thirds of lung cancer cases in the chapters could be linked directly to in-mine radiation and radon daughter exposure.[101] Of the nearly 5,000 Navajo men working in uranium mines, between 500 and 600 had died by 1990, with the same number expected to die before 2000.[102] Navajo uranium mill workers also had excessive levels of other diseases related to their occupations. A health study that followed more than 2,000 Navajo mill workers between 1940 and 1972 found that 1,500 men who had worked at seven different mills on the reservation had excessive levels of nonmalignant lung diseases, lung cancer, blood cancers, and chronic kidney diseases.[103] Numerous other studies have examined specific health abnormalities related to uranium exposure on Navajo land, though none has examined community-wide environmental health outcomes.[104] However, current studies do find that even low-level, long-term uranium exposure (for instance, to some of the 1,200 abandoned mines in the region) can lead to endocrine disruption, overproduction of estrogen mimics, and elevated rates of breast cancer.[105]

Nation residents and activist groups have witnessed numerous community environmental injustices in its chapters. Many Navajos still live in hogans built from tailings waste materials, have dealt with drinking water chronically contaminated by uranium exposure, and have shared homes with miners, millers, and their contaminated work clothes.[106] Observers note long-term negative psychosocial effects and other signs of social dislocation, and activists have formed several community support and information networks through the years to provide counseling and general support. [107]

PHOTO 2 This abandoned uranium mine, one of thousands estimated to exist in the region, is located just across Highway 141 from where the town of Uravan once stood. (Photo by Stephanie A. Malin)

Legislation, Compensation, and Remediation

Across the Colorado Plateau, the activism of uranium workers, their family members, and downwinders (that is, people exposed to radioactive fallout from test-site detonations between the 1940s and the 1960s) finally led to the 1990 passage of the federal Radiation Exposure Compensation Act (RECA), which included a 100-million-dollar provision for compensation but had little hard money attached to it.[108] RECA approved "compassionate payments" for deaths or injuries resulting from exposure to radioactive fallout or from work as a uranium miner. After meeting a variety of criteria, downwinders were awarded 50,000 dollars each for exposure. Uranium miners or their widows were awarded 100,000 each, as long as the widows could present marriage licenses and miners could prove they had been exposed to various amounts of radon. Awards also depended on a claimant's status as a smoker or a nonsmoker. Activists, however, argued that it was difficult for many miners or their widows—especially Navajos—to provide the necessary paperwork to gain compensation. In response, Congress amended RECA in 2000, simplifying its qualification requirements, offering coverage to uranium millers and haulers, and extending the eligible dates of employment to people who had worked through 1971. Uranium community residents who have not worked in the industry and are not considered downwinders are still excluded from compensation under RECA, despite widespread incidences of disease in their communities.

Environmental justice activism in others parts of the U.S. led to the passage of the Comprehensive Environmental, Response, Compensation, and Liability Act in 1980. For many uranium-contaminated sites, this opened the door to remediation, as did the Uranium Mill Tailings Remediation Act and the Uranium Tailings Radiation Control Act, both passed in 1978.[109] However, large-scale public health and environmental concerns in towns such as Monticello began to emerge only as remediation was under way.

Community Concerns and a Uranium Renaissance

Most uranium community residents know that the federal government and corporations have still not adequately addressed the legacies of uranium's previous booms, even though they are the very institutions that created the U.S. uranium market. However, public responses to boom-bust instability and environmental injustice vary drastically. As a nuclear renaissance looms, some activists have mobilized sites of resistance to renewed production and are fighting for recognition of uranium's legacies in the region. Activist groups such as the VMTE express frustration about the federal government's inadequate response to economic instability, environmental contamination, and negative health outcomes. These activists form a part of the Polanyian double movement, as they urge caution regarding the uranium industry, its privatization, and its regulatory capacities while fighting for increased social and environmental protections. Meanwhile, as I detail in subsequent chapters, other activists fight for local control over resource development, which they see as the most just use of their regional environment, and are mobilizing sites of acceptance to renewed uranium production. Some of these residents still deny that the industry is linked to environmental health risks, citing a 2007 study that found no statistically significant elevated rates of lung cancer among uranium workers in the Uravan region.[110] Although Union Carbide funded this study, which makes its conclusions suspect, these findings have been instrumental in creating reasonable doubt about the claims of uranium widows, workers, and other victims. These activists form a strand of what I call the triple movement, where people see markets as part of their social fabrics and support their privatization and increased extraction of related natural resources.

Support for renewed uranium development has long existed in pockets of the Colorado Plateau, yet not everyone shares that sentiment. Dependence on unstable uranium markets has created volatile economic conditions, with related natural resource dependence and spatial isolation structuring the legacies of environmental inequality. Communities have paid dearly for uranium development, losing their economic stability, their social cohesion, their environments, and their health, even as they have helped to create an American superpower. Today, as a third uranium boom looms, the region's overadaptation to natural resource extraction constrains their economic development choices. Material conditions such as spatial isolation and persistent poverty only accentuate these structural constraints. Economically and historically, residents are pressed to participate in the industry again, despite knowing its economic and environmental instabilities and risks. Importantly, privatization and reregulation make some residents believe that the industry might be able to avoid repeating the historical errors I have reviewed here. Amid such structural violence, all available choices reflect the social dislocation experienced by uranium community residents. But will renewed uranium production help them address the sociological foundations of that dislocation?

3

Lethal Legacies

Left in the Dust in Monticello, Utah

> The people across the fence—the wives, the children—do not qualify as "victims." They fall through the cracks.
> —Barbara Pipkin, Monticello resident

"You do what you have to do to provide for your family and put a roof over your head. My parents, their generation, they were so happy to have employment they didn't think there might be danger."[1] Those are the words of Fritz Pipkin, whom I introduced in chapter 1. A lifelong resident of Monticello, he has lived around and worked with uranium throughout his life. Fritz struggles with recurring lymphocytic leukemia and is now an activist with the group Victims of Mill Tailings Exposure (VMTE). His situation is familiar to many residents of uranium communities, including those who never worked in a uranium mine or mill. Today, even as uranium community residents hope for another industry boom, they struggle with the health and environmental legacies of uranium production.

Monticello is situated just east of Canyonlands National Park's sprawling red rock landscape. The high-elevation community lies in the shadow of the Abajo Mountains, but its rugged beauty is accompanied by spatial isolation and natural resource dependence. In the late 1930s, federal officials selected Monticello as the site of one of a handful of uranium and vanadium mills, reportedly because of its isolation, ready workforce, proximity to carnotite mines, and relatively abundant water supply.[2] Uniquely, the U.S. government owned the Monticello Mill throughout its operation, and locals still speculate that the mill's special federal status indicate that it was central in supplying yellowcake for the Manhattan Project and the bombs dropped on Japan during World War II. Little did they know that forty years later, four large uncovered tailings piles on the south end of town, adjacent to the southernmost neighborhoods in the community, would contain enough radioactive contamination to create two Superfund sites: not only the mill site itself but also community homes and infrastructure that had been constructed with materials from the radioactive tailings (see map 3). Those sites have been linked to ongoing, contested, and underaddressed environmental and health issues. By linking cancer clusters to long-term, community-wide uranium exposure, VMTE has united most of Monticello into a strong site of resistance to uranium production's social and environmental damages. Yet the federal government does not recognize the link between the illnesses and exposure, and for more than twenty years it has been contesting activists' claims. Thus, despite ambivalent support for industry renewal, Monticello residents and VMTE activists remain primarily concerned with issues of environmental justice and health.

Uranium's Political-Economic Power in Monticello

In my interviews, Monticello residents recollected strong local support for uranium development during World War II and the Cold War; and the archives of the *San Juan Record,* the weekly newspaper headquartered in Monticello, offer additional evidence that many saw it as both an economic reward and a patriotic endeavor. Front-page articles quoted President Eisenhower, a reminder to Monticellans that they were "engaged in a battle against world domination by communism."[3] An article on the American development of the A-bomb explained, "While chief protagonists in the Cold War were dueling with words, Uncle Sam could claim a theoretical 'touché' with the announcement that the U.S. is producing super A-bombs, substantially more powerful than the missiles that blasted Japan."[4] Uranium's strong patriotic and economic appeal had become so

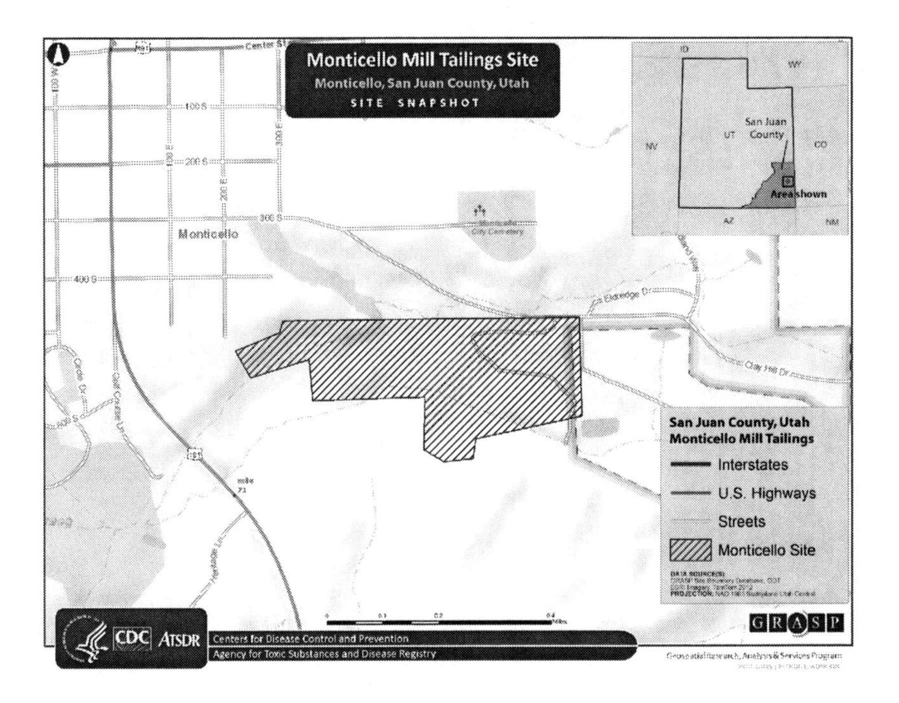

MAP 3 Monticello, Utah, and the remediated Monticello Mill site.
Source: U.S. Agency for Toxic Substances and Disease Registry, 2014.

engrained in community identity that they even featured it in the plotlines of serial stories that ran in the paper each week, while *San Juan Record* editorials praised San Juan County as "mak[ing] a greater contribution to the scientific world than any other spot in the universe."[5]

Accompanying patriotism and faith in atomic technology was a spirit of capitalism tied to uranium production. When the *San Juan Record* announced the construction of the Monticello Mill in 1941, the news meant "a great deal . . . not only [because of] the increased population during the months of building, but for all future time as the large forces of mill workers will continue their residence here."[6] The paper touted the mill as "the heart of our economy."[7] By 1949, the paper was reporting that "100 men are employed at the AEC mill."[8] Headlines framed uranium as a source of economic security for both the town and the region. Editors boasted that the "Monticello Uranium Boom Creates Million Dollar Building Program" and promoted "Grand Openings [That] Set the Pace for Expansion in U-Ore Center."[9]

As Fritz told me, some residents were so happy to have employment that they did not consider the possibility that the mill might pose significant long-term health or environmental risks. Still, at least some people were aware of possible trouble: in April 1945, the mill superintendent received a petition "signed by many of the townsfolk" and "asking for a check on the deadly sulfuric acid fumes which again spread through the community now that the mill is in operation."[10] Aside from this account, only one other story in the *San Juan Record* portrayed the mill negatively.[11] Yet Fritz recalled that the fumes from the mill's roaster stack gave the air a sulfuric "tang" and that they ate away at screen doors, car paint, and clothes hung out to dry.

Contested Illness and VMTE

By 1960, state-set prices for U.S. uranium ore had fallen, and the federal government ordered the closure of the Monticello Mill. The economic impact was immediate, and residents met the move with anger and protests. When the mill shut down, many other local businesses followed suit. Now the community had to reacclimate to a smaller tax base, reduced school enrollments, and a dwindling population, as is common during busts in natural resource–dependent towns.

For forty years the site, which was directly adjacent to neighborhoods at the south end of town, was left unremediated, the tailings piles uncovered. That site was home to approximately 100,000 cubic yards of contaminated materials, along with 2 million tons of tailings waste, contaminated soil, byproduct materials, and contaminated building materials. Dust from the tailings piles continued to blow into houses, and local children played on the tailings piles and swam in the creek that ran through them. Moreover, the Agency for Toxic Substances and Disease Registry has estimated that at least 156,000 cubic yards of radioactive tailings materials were used between the 1960s and 1990s in various Monticello construction projects, including as fill for open lands; backfill around water, sewer, and electrical lines and in basement foundations; sub-base for driveways, sidewalks, and concrete slabs; and sand mix in concrete, plaster, and mortar in community, home, and business construction.[12]

In 1989, responding to widespread contamination in Monticello and expanded federal legislation mandating uranium site remediation, the

Environmental Protection Agency designated two Superfund National Priority List sites in the community: the Monticello Mill Tailings Site and the Monticello Radioactively Contaminated Properties Site. A National Priorities List designation signals that contamination is so severe that remediation must occur as soon as possible and that all human activity on the site must be avoided. For the Monticello sites, remediation involved removing the four tailings piles, excavating 424 residential and business properties, locating contaminated materials in one secure repository south of town, and thoroughly destroying or burying all of them (see photo 3).[13]

During the cleanup, residents began noticing widespread health problems across the community that seemed more pernicious than previous issues associated with the 1945 smoke plume and the 1960s leukemia cluster. During interviews, activists and other residents told me that by the early 1990s they were hearing about new cases of cancer every week. They began connecting these illnesses to radioactive contamination from the mill and tailings piles on the site, though federal governmental institutions such as the Agency for Toxic Substances and Disease Registry did not conclude that the cancer clusters were related to uranium exposure, despite noting elevated rates of various cancers in the community. Thus, community members felt ignored by federal agencies and scientists, reporting to me that they did not receive satisfactory responses to their public health concerns.

When the VMTE mobilized in 1993, residents joined because they were tired of seeing neighbors suffer illnesses without explanation and felt they deserved more honesty from the federal government, particularly the Agency for Toxic Substances and Disease Registry and the Department of Energy, which owned the mill and acted as site custodian. A VMTE member told me, "Well, what we're all about is . . . to make the federal government make right the wrong they did to the community." Another said, "We want something done by the government because right now [cancer] just keeps happening." Yet another said that when people start thanking activists for their work, "you don't have a choice; . . . you have to keep going on, and there are times it is discouraging . . . but we *have* to go on." All of the members concur. They are involved because they feel the federal government should implement basic social safety nets: that is, acknowledge their illnesses, accept responsibility for what the VMTE contends are direct effects of uranium exposure attributable to the Monticello Mill,

PHOTO 3 Radioactive waste materials removed during remediation of Monticello's two Superfund sites were stored just south of town in these repositories, pictured in the foreground of the photo. The former mill site lies just north of the repository, in the middle of the photo. The town of Monticello appears at the top of the photo. (Photo by Stephanie A. Malin)

and make sure that sick residents can find and afford healthcare despite their spatial isolation and economic vulnerability.

Though membership fluctuates, seven people are the core of the group: two retired homemakers, a machine-shop owner and city councilman, a liquor-store manager, a retired state employee and city councilman, a county economic development officer, and a former miner and miller. Given their spatial isolation, community activists must work hard to publicize their activism. They seek out interviews with media sources throughout Utah, they have constructed a website, and they organize annual events such as the Monticello VMTE Cancer Awareness Walk and Celebration of Life.[14] As advocates for the community, activists are also involved in mapping the incidence of disease (a practice known as popular epidemiology) and lobbying for appropriations to cover medical expenses and bring in temporary medical facilities.[15] In 1993, 2005, and 2007, they designed and conducted community health surveys both in person and via the *San Juan Record*. So far, VMTE activists have recorded nearly six hundred cancer cases (twenty-six of them leukemia) and more than a hundred cases of respiratory and other health complications such as universal allergies—all in a town of 1,958 people.[16] The 2007 survey, which was conducted with assistance from the Utah Department of Health, firmly established that town had elevated rates of certain cancers, including stomach cancer.[17] Members have even scraped together funds to fly to Washington, D.C. to lobby Congress for appropriations. Though they had temporary success in 2007 and 2008, the ongoing recession has reduced federal funding for community cancer screening and medical coverage. Still, the VMTE has worked with state senators to secure annual financial support for temporary screening clinics in the community.

VMTE activists have four main goals, which are overwhelmingly supported by Monticello residents. First, they want a cancer screening and treatment facility in Monticello. Currently, residents must travel to Durango or Grand Junction, Colorado (106 and 167 miles away, respectively), or St. George, Utah (394 miles away), to receive adequate care. According to survey data, most respondents believe that a local clinic would increase the likelihood of early cancer detection and make treatment more affordable and comfortable. Second, members want a federally funded trust to pay for the medical expenses of every Monticellan in need. Currently, paying for treatment is what one VMTE member has called a "spendy little adventure." According to survey data, hundreds of residents have taken out second mortgages, declared bankruptcy, or endured other financial strains when trying to pay for cancer treatment.[18] My own fieldwork verifies this situation: one couple I interviewed took out two mortgages on their home and spent 70,000 dollars on the husband's leukemia treatments. As one VMTE member told me, "What we expect . . . [is that] the government is going to pay the bill [for] early detection and prevention and treatment right here, because it was negligent and it was their mess." Third, the VMTE wants the federal government to recognize and thus legitimize residents' contested illnesses. A simple apology would mean the world to many Monticellans, but this could be risky and expensive for the government, especially if compensation under RECA were extended to uranium community members. Finally, activists request a dose reconstruction study, which would publicly establish what types of chemicals and processing techniques were used at Monticello Mill during its operation. Such knowledge would help interested parties come to more definitive conclusions about the relationship between residents' illnesses and exposure to the uranium mill's pollutants.

PHOTO 4 The Victims of Mill Tailings Exposure have worked for decades to fund a Monticello Mill Memorial Site atop the remediated mill grounds, complete with walking paths and a timeline of uranium legacies. (Photo by Stephanie A. Malin)

Distrust in the Federal Government

The findings of the Agency for Toxic Substances and Disease Registry are the basis for the government's neutral stance on causal links between uranium exposure and elevated rates of disease in Monticello. Yet those findings themselves are contradictory. The agency's 1997 report identified and recorded multiple pathways through which residents may have been exposed to radioactive uranium and found evidence of severe environmental contamination and negative public health consequences.[19] It also asserted that cancer rates in Monticello exhibit unusual patterns, noting a 395-percent increase in men's deaths linked to cancer of the trachea, bronchus, and lung pleura between 1950 and 1980. Among Caucasian women, rates in breast cancer jumped 287 percent during this same period. These increases are striking, especially when one considers that Utah has the lowest rate of cancer in the nation.[20] In addition, age-adjusted rates for renal failure in both men and women in San

Juan County were higher during this period than they were in the rest of Utah, as were rates of birth defects.

Nonetheless, because of the study population's small numbers, the 1997 report concluded that standard errors for each investigation were high, meaning that little could be statistically shown about cancer rates in Monticello. Thus, instead of connecting elevated cancer rates to uranium exposure and legitimizing residents' experiences, the official document on the mill's public health effects in the community called for further monitoring and research. It made surprisingly few recommendations about public health concerns related to cancer, despite the agency's responsibility for monitoring those very outcomes in community populations. Nor, say my interviewees, did agency officials respond adequately to local concerns when they met with community members.[21]

The people of Monticello are chronically frustrated by government inaction. As one VMTE activist told me, "[the community] questioned the government many times and [got] the same response: there's not a problem down there. But cancer is in every neighborhood, every other house." During my fieldwork, Walter, the VMTE's founder, gave me a stack of formerly classified documents, VMTE survey results, and assorted health studies of Monticello. Among them, I found an independent assessment of the 1997 Agency for Toxic Substances and Disease Registry report. It was conducted by the Peer Review Committee of the Consortium for Risk Evaluation with Stakeholder Participation, comprised of nineteen independent scientists from various U.S. universities and health institutes. The assessment notes the "incomplete" nature of the agency's report and expresses concern about the agency's failure to equip Monticellans with knowledge of events in their community and how their health was or would be affected. According to the committee, "the Public Health Action Plan presented [in the agency's report] is not a clear, concise, or orderly delineation of activities that will be pursued to assess levels of current or past exposures, or to prevent the occurrence (and, failing that, progression) of diseases or dysfunction which may result from exposure to the radioactive tailings and other contaminants. The plan is so vague that it is difficult to determine whether (when translated into action), it will interrupt exposure or otherwise reduce the public health risk."[22]

While agency scientists were conducting their initial research, Monticello residents were using popular epidemiological methods to create their own health study. In 1993, two local women distributed a survey that included questions about respondents' smoking habits. Boston University School of Public Health scientists helped the women analyze the data, which revealed odds ratios for contracting cancer in the town. The results, though not from a random or a scientifically drawn sample, indicated that even nonsmokers in high radon areas had an elevated chance of getting lung cancer.[23] The study's results were shared at a town meeting.[24]

As such information was publicized, many residents began questioning how accurately the Agency for Toxic Substances and Disease Registry had assessed their environmental health experiences. *San Juan Record* reporter Lee Bennett's front-page article assessed community-wide dissatisfaction with the report: "Many residents and former mill workers hoped the public health assessment would provide insight regarding the relationship between health problems and the mill operations. However, the report may generate more concerns than answers."[25] Bennett criticized it for "stop[ping] short of saying local illness is caused by contaminants from the mill site . . . [despite] finding several health risks to Monticello residents." He worried as well that "the current multi-million dollar clean-up effort does not include infrastructure issues" and highlighted the report's inaccessible language, explaining that "pathway is a fancy way of saying that scientists have confirmed that radioactive or chemical toxins (including uranium and vanadium) are consumed or absorbed by Monticello residents."[26]

Bennett's criticisms betray an underlying distrust in government scientists—a distrust that has since become common throughout the community. At public meetings, VMTE members and other residents have repeatedly asked scientists to provide more complete and transparent assessments of chemicals used at the mill as well as levels of community exposure to those contaminants. But for more than twenty years, those requests have also gone unanswered. At a May 2006 meeting in Monticello, Cliff, a VMTE member and retired state employee, asked agency officials to do a dose reconstruction study. He was told that the agency already has that information. Yet as Cliff told me, "[The people of Monticello] don't know what went into that mill, we don't know what came out of the mill, and we don't know what was trapped in the vegetables . . . [that] people ate out of the gardens." Jayne, a VMTE member, the wife of a uranium miller, and a lifelong Monticello resident, said that as soon as she saw government employees cleaning sites in "moonsuits," she needed to know exactly what came out of that mill. She did not feel that the agency report had answered that question or that officials had adequately responded to it at public meetings.

VMTE members continue to criticize the federal government's lack of transparency. Jayne's husband, Herman, a bone cancer victim, told me that he still wanted "to know what was coming out of that smokestack and what was seeping into the ground and the air from those tailings piles for all those years." Barbara Pipkin observed, "So many of those records are classified or disappeared, or are suddenly lost or whatever. It's become really hard to find out what came out of that mill exactly. So I guess at times it makes us angry." Claire, a VMTE member who is also an economic development officer for San Juan County explained, "For the first health study . . . records weren't available, they hadn't been declassified. Even now, we don't know what came out of there." Walter, founder of the VMTE and a local machine-shop owner, said that his desire for a dose reconstruction study was triggered by the cryptic comments of Department of Energy agents during a site visit in Monticello:

> The DOE gentleman says, "We know you were contaminated." Okay, we've been contaminated for fifty years; tell us what we were contaminated with. They know we were contaminated for fifty years, but the level of cancer has not increased. Give me a break: that doesn't make sense. Why did they come here to clean it up, if it wasn't dangerous? . . . I think if they would address the issue and tell us what we were exposed to, those reconstructions would tell us it was contaminated. [A dose reconstruction study] tells if there was a problem to watch out for, it tells if there wasn't a problem, and then lets us go on our way. Simple as that: tell us what we were exposed to.

Fritz expressed similar concerns: "A fellow from the DOE, he said, 'Yes, there was contamination,' but they won't tell us to what extent. And it wasn't just low-grade uranium or wastes. That is the problem: it's all of it. I mean, all those chemicals that they used in the process—the hydrochloric acid, sulfuric acid, the chemicals to break it down. . . . I mean, [the mill] set there from 1941 till 2000 when they finally got it cleaned up. And now they've given compensation to mill workers after being around the mill for two years. But this thing set there for dang near sixty years."

Residents spoke of other uranium communities, such as Uravan, just fifty miles away, that were so contaminated by uranium production that they had to be demolished. They wonder if the severity of Monticello's own contamination—two Superfund sites within the town limits, the widespread use of tailings to build community infrastructure—means that it cannot ever be completely remediated. Though she loves her hometown and her neighbors, Jayne told me that "the town would have been better off if they just would have relocated us somewhere else, somewhere safer." She and Herman still live on the edge of the old mill property, and she said that no camping or other long-term use is allowed on the site, even after remediation. She suspects that officials still harbor reservations about the property's safety and wondered sarcastically "if the contamination just magically stops before it reaches my home or the homes of the rest of the folks in this town."

For years, articles and advertisements in the *San Juan Record* have echoed Jayne's concerns. Newspaper reports mentioned AEC-owned homes, once located on the mill site, that were sold to townspeople shortly after the mill closed.[27] Ads published between the mill's closure and Superfund remediation in the 1990s demonstrated the way in which site infrastructure would be used throughout town—for instance, "City Will Utilize Mill Tanks to Increase Water Storage Capacity."[28] If we recall that the government knew at the time that radioactive contamination was present in site materials such as those tanks, we can understand why residents are still concerned about the thoroughness of remediation. Tailings debris used in community construction has been another ongoing issue: the federal government has still not accounted for all 156,000 tons of that material. Instead, during the cleanup period, it regularly sought residents' aid via newspaper announcements such as "The DOE requests your assistance in locating materials or scrap that may have been removed from the Monticello Mill site and taken to properties in and around Monticello, Utah." The ads warned residents that "if your property is contaminated, there is a potential health hazard from exposure to radon gas or contaminated materials."[29]

Several documents declassified during the 1990s reinforced residents' distrust of the federal government—among them the 1940s Department of Defense memo (discussed in chapter 2) that instructed Nevada Test Site scientists to wait to detonate test bombs until atmospheric winds could blow the fallout over sparsely populated southern Utah and northern Arizona.[30] This memo's language deeply affected VMTE members' perceptions of the federal government. As Fritz said, "We've seen the document that came out years ago when they were doing the testing in Nevada for the atomic bomb. It said that all we were was 'a low-use segment of the population.' . . . It was the same process with this mill." Other VMTE members echoed his sentiments, with one noting, "I think [that] statement . . . makes our group madder than anything." Although government reaction to the memo has been limited, documentarian Carole Gallagher has recorded a strikingly callous remark by Barbara Yoerg, a public affairs officer at the Department of Energy's Las Vegas offices whom Gallagher interviewed in the late 1980s about radioactive tests and the "low-use segment" memo. When asked about the practice of waiting until the wind was over southern Utah to run the tests, Yoerg replied, "Those people in Utah don't give a shit about radiation."[31]

VMTE members have gained access to declassified papers about Duncan Holaday's Public Health Services studies of the Colorado Plateau (discussed in chapter 2), which the government had forbidden him to make public.[32] They have also obtained a mid-1980s Bureau of Land Management report that openly discussed radioactive contaminants on the mill site and warned federal employees against inhabiting it. Explaining the bureau's decision not to build a new office on the site, the report states, "The DOE did a report on gamma radiation and radon data [of mill land] and advised us by letter . . . that there was no 'significant radiological problem associated with construction in the area near the old AEC lab building.' [But] by verbal communication [the Department of Energy] said, 'don't do it!'"[33] According to the report, a department official verbally admitted to a bureau staff member that the government knew about many "hot spots" on and around the site.[34] It even quoted the person as saying that the "property should be clean before it is offered to the public." In other words, the government allowed Monticellans to live in close proximity to the mill site for decades before and after the report was written, all the while knowing that "the property is a safety hazard that we should not expose our [government] employees to and should abandon as soon as possible."[35] Yet Monticello's children continued to play on tailings piles and swim in Montezuma Creek; their parents and town officials continued to use tailings materials from the site in various construction projects around town.

Why, asked VMTE members, weren't the townspeople allowed to make equally informed decisions? According to the group's surveys and my fieldwork, failure to protect Monticello's citizens was the single biggest contributor to residents' loss of trust in the federal government. They felt betrayed. By extension, this loss of trust tainted everything else that they had worked for during the war years and tarnished the community's role in establishing America as a global superpower. Fritz told me:

> My father worked [at the mill]; I was raised exactly a block down the street. And when I was four or five years old, I'd see him come down the street in his old pickup, and I would step up on the running board, and he would hold me on and go home. And he came home with his work clothes on at lunch and he came home with them at night. And dust and all this stuff was on their clothing. They brought it home, washed it with everybody else's clothing. *And nothing was ever said.* We had a cow we kept right next to the mill, and we drank the milk for years, we played in the tailings piles—*they never chased us out.* As boys, . . . instead of coming home to get a drink, we would drink out of the crick that ran through those tailings. We spent endless hours down there playing, and they encouraged our parents to haul off the waste product. They . . . used it for road base and mortar and sidewalks and children's sandboxes. And when [the government workers] came to clean up, . . . they [were] completely masked. They won't let them on or off [the site] without monitoring or hosing them down.

An article in the *San Juan Record* echoed Fritz's reaction: "People supported the government, took up their positions, and did their best to assure our country remained a winner. And then people got sick and wondered what the government didn't tell them. Government officials, after studying the situation, say the sickness is just an odd coincidence. Are we talking about . . . Love Canal in New York State? No, we are talking about radiation and toxic chemicals in Monticello. We're talking about the workers at the Monticello Uranium Mill, their families, their community."[36] According to Claire, "The U.S. government knew exactly what was going on. . . . While they were here, they were dumping 2,500 pounds of particulates into the air every day, and all of Monticello is in the plume because the mill is just down there. Then they left and put a barbed-wire fence around it to keep cows out. They didn't do anything to keep people out."

There is a widespread community belief that the region was a national sacrifice zone, and perhaps still is. In an open letter to President George W. Bush, resident Joe Torres, then dying of lung cancer, expressed fierce anger at the federal government's deceit:

> You might ask how this all came about. . . . Many years ago the Government built a mill to process uranium and allowed the community of Monticello, Utah, to be exposed to unhealthy levels of radiation. Today, everyone in our

> little community has been touched by cancer. Cancer does not just kill one, it kills the family. . . . The disaster came in very slowly, as we quietly slept, as the children played outside, as we worked. There was no warning. It snuck into our homes, permeating the very air we breathed and the clothes that we wore. There wasn't a lot of fuss made, . . . daily life continued in Monticello. We didn't know what was to come. Now, everyday another resident is informed they have cancer. . . . My time here is now very short. I am tired and would like to know that maybe some of what I do now might make it so that other folks will not have to wait and be forgotten like I have been. It is hard to fight cancer and fight the government.[37]

The VMTE recognizes that the community's powerlessness is linked to its rural isolation. In fact, the town's isolation is a main factor in the long mobilization of this site of resistance. During my fieldwork, VMTE members repeatedly lamented the town's lack of power due to its rural location and consequent lack of potential activist networks, political support, and media attention. At the end of one of my early interviews, one of the organization's founding members asked me for any help I could offer, wondering aloud, "How do you fight the federal government?" Later another member said, "I don't think [government officials] care because it's a low-use segment of the population. . . . We don't have the power to vote them in or out." Many commented, "We need Erin Brockovich to come down here."[38] One declared, "I don't think the government was right to do what they did, and it's our responsibility to make them do what they are supposed to, even though we're a tiny little town in the middle of nowhere." During a meeting in early 2007, a field officer for a Utah state senator told me, "Monticello is worse than Love Canal but it gets ignored because of its isolated geographical location." Isolation has limited Monticello's access to nearby sites of social networking and scientific collaboration, thus exacerbating the legacies of environmental injustice in the community and highlighting their spatial and structural disadvantages. For example, until 2006, VMTE members could not afford to regularly travel to Salt Lake City to meet with representatives from HEAL Utah, the state's antinuclear lobby. While the group has been resourceful—accumulating government documents, evidence of exposure, and ties with extralocal environmental justice nonprofits and organizations—it has needed twenty years to form these distant ties and fortify itself as a site of resistance.

"It Slowly Bleeds You to Death": Social Dislocation and Worldview Shifts

Like other people living in contaminated towns, Monticello residents and their extended families continue to undergo profound changes in their worldviews.[39] As they have navigated through the experiences of toxic exposure and discovered that the burden of proof often falls on affected communities and residents rather than the entity that created the contamination, their notions of justice have also shifted. Much of the life that they once took for granted—good health, honesty and transparency from government and local leaders, liveliness and happiness in the community, faith in science and the American Dream—have vanished. *Worldview shifts* is my term, not necessarily theirs. But no matter their word choice, the people whom I interviewed told me that their lives have been irreversibly changed.

Monticellans' anxieties are not limited to their own illnesses. Because uranium and other radioactive components have such long half-lives, the health of future generations will also be affected. This fear and uncertainty reinforces the social dislocation that people feel when scientists or officials tell them they are imagining connections between their health outcomes, the Monticello Mill, and its long-abandoned, contaminated site. Residents see patterns and health problems every day; and their expressions of frustration are symptoms of how they now view the world, their town, their friends, and their health.

For VMTE members, contesting the public health findings I have discussed is an arduous process, almost a full-time job. At the same time, those activists who are dealing with cancer are juggling a set of stressors that includes financial insecurity. Barbara told me, "Fritz is alive and able to work, to hold down a job. We're some of the lucky ones. There's so many more that [illness] has devastated. . . . It slowly bleeds you to death." Discussing the community's substantial health problems in the context of reduced federal social safety nets, Walter said, "I know the people of this town who suffer from cancer have to go through an incredible expense. . . . It breaks them, it breaks their families."

According to Claire, contemporary environmental health outcomes in uranium communities are *not* the result of governmental ignorance in the 1940s and 1950s: "The pervasive attitude that back in the forties [the government] didn't know what all the results were going to be of the tests and the radiation: I don't believe that. I think they did know." Joe Torres echoed Claire's suspicions in his open letter to the president: "The politicians can now stand up and tell everyone that they feel our pain and have heard our voices and did something about it. But they didn't. . . . This kind of trickery was completely new to us. You probably think we are all simple, and perhaps we are. We believe in simple things, including if a man says he is going to help you, you can bet he will. . . . To be near the end with no relief from the government has saddened me very much."[40]

Hopelessness has become widespread in the community. As Jayne lamented, "so many people in this town are just so sad right now." VMTE members told me that their group stays small primarily because most residents are tired of fighting. One activist said, "There are a bunch of them who have been beat up so badly by the federal government that they just wish [the problem] would go away." At the same time, residents are preoccupied with the idea that one day they, too, will wake up with cancer. According to a VMTE member, "people will tell me, and I feel it too, that every time they feel a tickle in their throat or have a sneezing or coughing fit . . . the first thing we think of is 'Do I have cancer?'. This is no way to live." Walter said, "I'm going to get cancer; it's just a matter of when. . . . I think most of the community feels that way." Even the fiercest community activists fear they will soon lose their ambition for activism. Walter frequently told me, "I'm just tired and about to quit this thing." As defeat overwhelms the residents, VMTE activists have more and more trouble persuading people to complete their health surveys. Barbara explained, "We have sat with these people and cried about our town, and about our losses," but people are too consumed with family illnesses and medical bills, too reluctant to relive painful memories, or too skeptical about challenging the federal government to return their surveys or join the VMTE's site of resistance.

"It's Just the Chance You Take": Ambivalent Support for Industry Renewal

Today, many Monticellans trust the corporations that currently dominate privatized uranium markets more than they trust the federal government. According to some VMTE members, the privatized industry might operate under more stringent regulations than the government did and create a safer domestic energy economy. Yet even though these activists ambivalently support renewed uranium production partly because of persistent poverty, their community's environmental justice and health concerns overshadow the industry's economic allure. In Monticello, sites of acceptance have not yet mobilized as they have in Nucla and Naturita. Residents remain concerned about unequal exposure to environmental toxins, access to information, and recognition of local illnesses.

Still, since 2007, renewed uranium production has emerged as a real and familiar option for regional economic revitalization. The proposed Piñon Ridge Uranium Mill would be sited sixty miles east of town, and the White Mesa Mill—the only operational uranium mill in the United States—is just twenty miles south in Blanding, Utah. Residents' ambivalent support for production hinges on cautious faith in industry regulations, which are enforced by corporations and the state of Utah rather than the federal government. Economic need and the lack of federally provided social safety nets also facilitate this trust. Nancy, a VMTE member and a nurse who has since lost her life to cancer, told me that she has faith in regulations because now they actually exist: "We do know there's been enough changes made from the time [Monticello Mill] operated in the 1940s. I mean, [mill workers] had no protection or anything. So certainly [the industry has] learned from that and realized that people aren't guinea pigs and that they have some worth. . . . You know, there's always something that could happen, but . . . I am for uranium exploration and milling. I think that's the way of the future; I think that it's necessary. It's something that's rich in this area."

Walter believes that regulations have improved thanks to private industry oversight, but he recognizes his own economic dependence on the industry: "[The Monticello Mill] was government-controlled, and they were on a time schedule for the war effort. Get it done, get it done, I don't care what it costs or who it hurts. Then they walked away. . . . They were not regulated. . . . Piñon Ridge, . . . that'll be a private mill and there will be regulations. It will be the newest facility of all the mills, so it will have all the improvements that everyone's learning from. So personally, I'm supportive of uranium mining because I'm in that business. But it needs to be done safely—mine and mill, safe."

Three considerations underscore neoliberalism's contemporary hegemonic power in uranium communities: new faith in devolved governance, strong disembedded economic markets amid persistent poverty, and pride in uranium's role in renewable energy. All remain at the core of VMTE activists' tentative support for industry renewal. Some members, such as Nancy, also believe that regulations will be better if uranium communities are equipped to oversee corporate performance in mines and mills: "I think probably the best thing you can do as a community is to make a relationship with the people who run the mill or the mine or whatever." Regardless of their positions on corporate transparency, however, the activists see uranium as a central component of their local economy. Walter, who owns a machine shop, acknowledged that his faith in regulations emerges from economic constraint, which conflicts with his distrust of the government: "If [Piñon Ridge] is opened over there, the local mines will open up. When the mines open up, they want more equipment.... [But] it's kind of a catch-22, if you can't trust your government, and a lot of us question our government. But you've got to trust them, I guess, as far as regulations go." Likewise, Nancy said, "There are so many things economically that we have to have help with. There are only so many things that are right here. So I think anything that's new economically, people are going to be in favor of."

Most VMTE activists believe that nuclear power will provide cleaner energy than other sources do. This adds to the appeal of industry renewal, and even makes it seem patriotic. As Barbara, a fierce critic of the government, observed, "You consider maybe they're not building the atom bomb to free the world but they're trying to free the world of the chokehold of foreign oil." Yet residents are still acutely aware of risk and the role of structural violence. Fritz noted the relationship between economic need and constrained community choice about economic development: "You can see how it works, trickles down. Like, ... it's [Walter's] life and [his] profession to build these mine buggies, ... and his boy works in the mill [in Blanding]. I used to work in the mine. And yet every one of us kind of has a little wonder, even Walter, out there working with those dirty buggies that they get back in to repair. I bet you every one of them out there kind of wonders once in a while, 'Is there some kind of contamination on these buggies?' It's just the chance you take." Barbara said, "All along, we've been told that 'It's safe, it's safe, it's safe.' And that all we can do is go on their word that it's safe. Who knows? Nobody knows what *safe* is. It's a risk we take every day."

As Monticello's case illustrates, the federal government has still not adequately addressed the legacies of previous booms. But with prices for ore and refined uranium increasing from less than ten dollars per pound in

2002 to more than fifty dollars per pound today, Monticello and other uranium communities see promise in domestic nuclear energy development. Despite environmental injustices and ambivalent trust in corporations and regulators, residents—even those who have created sites of resistance—feel economically constrained to accept uranium renewal, albeit tentatively.

Before nuclear power is framed and funded as a renewable energy source offering social sustainability, we must reconcile the underaddressed environmental, health, and social legacies in these communities. Further, we must carefully consider the structural disadvantages and disempowering implications of a context in which support for the industry emerges from economic dependence and poverty so persistent that it constrains the transformative potential of social activism. Activist organizations such as the VMTE work toward goals (for example, access to nearby medical facilities and healthcare) that reflect the nation's reduced public spending and lack of social safety nets, especially in rural communities. Even sites of resistance such as the VMTE's are less empowered to work toward transformative changes when their local economies have the infrastructure and skills to facilitate renewed uranium development and when neoliberalized structures and norms discourage collective solutions. As such, even in this site of resistance created by the VMTE and motivated by notions of environmental justice and health, structural violence is done to activists' goals and tactics.

Further, Monticellans' health issues are troubling because they continue with full knowledge of the federal government and have been confirmed by public institutions such as the Utah Department of Health.[41] If rural communities across the Colorado Plateau are to be transformed again into centers of uranium mining and milling, then residents must have ample access to screening facilities and high-quality healthcare. In addition, the EPA and the Agency for Toxic Substances and Disease Registry must be held accountable for citizen and community well-being. These institutions must be more transparent in their findings, more proactive in their conclusions, and more responsive to the public: they must serve as ordinary citizens' recourse against other powerful institutions. The Agency for Toxic Substances and Disease Registry has already acknowledged the need for improvement by initiating a program called the National Conversation on Public Health and Chemical Exposures, which aims to "develop an Action Agenda with clear, achievable recommendations to help government agencies and other organizations strengthen their efforts to protect the public from harmful chemical exposures."[42]

Yet even with these and other improvements, should we still consider pursuing uranium mining and milling or supporting the nuclear fuel cycle as a sustainable technology? Are forms of social dislocation related to uranium production merely historical and tied only to the industry's legacies? Recent research shows that very low levels of uranium exposure may disrupt endocrine systems and hormonal levels, thus leading to diseases such as breast cancer.[43] New findings report that women living within five miles of a uranium processing plant have elevated rates of breast cancer and other diseases.[44] Uranium may be too risky even in our risk-prone society.

4

The Piñon Ridge Uranium Mill

A Transnational Corporation Comes Home

> Safety, health and environmental responsibility are also of paramount importance to the Company. Energy Fuels is committed to the operation of its facilities in a manner that puts the safety of its workers, its contractors, its community, the environment and the principles of sustainable development above all else. . . . We believe we are positioned to respond to market conditions and achieve sustainable and responsible long-term uranium production as the World's demand for clean, affordable nuclear energy continues to grow.
> —Energy Fuels Resources

The Piñon Ridge Uranium Mill is the first such mill to be permitted in the United States since the end of the Cold War. Across the Colorado Plateau people are mobilizing sites of acceptance and sites of resistance to renewed uranium production, and most are focusing on Energy Fuels' proposed

Piñon Ridge Uranium Mill. Since acquiring Denison Mines Corporation in May 2012, Energy Fuels has become the largest conventional uranium and vanadium producer in the United States.[1] The corporation has consolidated control over American uranium markets, counting among its assets the nation's only operational uranium mill (White Mesa) as well as the land and permits for Piñon Ridge and several mines in Utah, Colorado, Arizona, and Wyoming.[2]

Technically, Energy Fuels Resources, Inc., is a subsidiary of the transnational corporation Energy Fuels, Inc. Yet many residents of Paradox Valley think of it as a local company. George Glasier, the company's founder and former CEO, lives on a ranch outside Nucla; and he and his wife have had a long, close relationship with communities in the region. In addition, the company has developed a strong local presence by holding public meetings, opening local branch offices, and employing residents to collect baseline air and water samples before mill construction. Support for the company and its activities is not a case of what is sometimes called "AstroTurf activism": Energy Fuels does not *create* local activism in support of Piñon Ridge but carefully cultivates grassroots mobilization. It uses its largely favorable reputation among Paradox Valley residents, an active local presence, and its extensive uranium assets to enhance what is already an atmosphere of strong support for renewed production. Simultaneously, however, the corporation's activities shrink the social and political space available for people to critically explore or resist renewal. Instead, the company capitalizes on residents' trust, their identification with the industry, and class-based tensions with wealthier communities such as Telluride to strengthen active sites of acceptance.

At the core of the conflict between sites of acceptance and sites of resistance are strikingly different perceptions of environmental justice in general and of Energy Fuels, the Piñon Ridge Mill, and uranium production renewal in particular. While supporters believe that renewal will reduce persistent poverty, enliven communities, and be environmentally safe, opponents focus on risks and legacies associated with the industry and their distrust of Energy Fuels. Even as policymakers increasingly frame nuclear power as a renewable energy source, these local disagreements highlight inconsistencies in that idealistic characterization. In reality, Energy Fuels' attempts to revitalize the industry stoke longstanding class conflicts and block possible cooperation in community development or alternative energy innovation. Opposition activists leading sites of resistance have received death threats and been harassed by pursuing drivers on desolate state highways; supporters have been labeled as ignorant or seen as victims of the Stockholm syndrome. Clearly, material conditions including natural resource dependence, persistent poverty, spatial isolation, a strategic corporation, and familiarity with the uranium industry have created divergent notions of environmental justice and activism.

A Symbol of Renewal

The 880-acre site of the proposed Piñon Ridge Uranium Mill nestles along State Highway 90 in the western end of Colorado's Montrose County (see map 1). The area is known as Paradox Valley, a beautiful red-rock landscape interspersed with abundant green meadows between towering rock walls. The arid valley stands about 5,500 feet above sea level, with the Dolores River running perpendicular to it about seven miles west of the mill site. The valley's name originated from that river's paradoxical path, created when salt-dome caves collapsed to form the landforms. Montrose County covers roughly 2,200 square miles, about 68 percent of which is public land managed by federal agencies.[3] Major land uses include mining, agriculture, and recreational activities on a wide variety of those public lands.[4] Until the county commission rezoned the mill site for industrial purposes, most of Paradox Valley was zoned and used as residential or agricultural land, mostly for grazing. The communities of Nucla and Naturita lie closest to the proposed mill site, which is about sixty miles east of Monticello and sixty miles west of Telluride.

As the first uranium mill sited in the United States in nearly thirty years, Piñon Ridge has captured the public imagination while stirring drastically divergent responses to nuclear renaissance. For supporters, the mill symbolizes social and economic revitalization, a chance to patriotically combat climate change while returning to their roots in the uranium industry and increasing their prosperity. For opponents, the facility symbolizes underaddressed environmental legacies and risks, an unstable global market prone to boom-bust cycles, economic dependence, and threats to their own livelihoods, which often involve tourism, recreation, or sustainable agriculture. This division gives Energy Fuels and pro-nuclear policymakers an unusual advantage. While they are used to overcoming sites of resistance to industrial facilities, the presence of local supporters mobilizing sites of acceptance

PHOTO 5 To approve the permit for the Piñon Ridge Mill, Montrose County commissioners voted to rezone Energy Fuels' land from agricultural to industrial use. If built, the mill will be surrounded by land used by generations of farmers and ranchers. (Photo by Matthew Kazy)

lends legitimacy to the project. Energy analysts and activists have become interested in company claims that the mill will be the "most environmentally friendly . . . in the world."[5] The company has also spent time showcasing the proposed facility's technological superiority, its own "commitment to transparency," and its positive working relationships with Colorado regulators.[6]

Permitting the Mill

Since 2007, when it announced plans to construct the Piñon Ridge Mill, Energy Fuels has established itself as a uranium stronghold on the Colorado Plateau. It has actively expanded and renovated mining operations in the region, which it says would provide about 70 percent of the uranium processed at the proposed mill.[7] Yet fulfilling the Colorado Department of Public Health and Environment's licensure requirements has not been easy. Colorado is an Agreement State, meaning that it has proved to the Nuclear Regulatory Commission that its enforcement capabilities equal or surpass the federal commission's regulations for all uranium- or nuclear-related activities in Colorado. Thus, the department, not the U.S. government, is responsible for granting final licensure for facilities such as Piñon Ridge Mill; it also oversees all monitoring to ensure regulatory compliance.

For Energy Fuels, this process has been neither short nor sweet, despite the corporation's close working relationship with the department. Each step of the facility's siting and construction has been intensely scrutinized and has required years of work. To complete its application for licensure, Energy Fuels had to submit fifteen volumes of reports, plans, and preliminary investigations, which the department approved in December 2009.[8] But Energy Fuels' radioactive materials license application for Piñon Ridge did not receive final approval until January 2011.

Nonetheless, the corporation has not yet broken ground at the mill site. Construction has been delayed by repeated legal challenges to not only the permit but the corporation's and the department's approaches to eliciting public responses to the facility. In 2012, responding to lawsuits by the Sheep Mountain Alliance and other anti-mill organizations, Denver District Court judge John McMullen set aside the radioactive materials license, ruling that Energy Fuels had not held adequate public hearings, even though it had hosted multiple public meetings. Concurring, the Nuclear Regulatory Commission ruled that Colorado had to allow more open public participation to meet regulatory standards by holding a formal hearing in which concerned citizens could formally question company and regulatory officials.

In November 2012, the state held this formal public hearing; and in January 2013, Judge Richard Dana, who had presided over that hearing, found that Energy Fuels had followed protocol in agreeing to host it. In turn, the department once again took the corporation's licensure application under review. In late April 2013, it recommended that the state reissue a radioactive materials license to Energy Fuels, once again giving the company permission to construct the mill. In an official statement, department chief Chris Urbina said, "From the beginning, we have listened carefully to the public and worked with Energy Fuels to minimize risks to public health and the environment. . . . [We enforce] strict environmental regulations [that] far exceed those in place when the last such mill was constructed more than 25 years ago."[9] Yet this is undoubtedly a short-term victory for Energy Fuels, and contentious public debate will continue. The industry's boom-bust patterns will also affect the fate of the Piñon Ridge Uranium Mill; if legal challenges delay mill construction for years to come, global uranium prices may once again decline and discourage construction of the facility.

A Peek Inside

A few houses and farms dot the landscape near the Piñon Ridge Mill's proposed site, but the facility would operate far from major population centers; 90 percent of the land within five miles of the site remains uninhabited by people.[10] Plans include the seventeen-acre mill, about ninety acres devoted to tailings ponds (which reduce airborne radioactivity of tailings piles by immersing the uranium ore material in liquid), a forty-acre evaporation pond, a six-acre storage pad for uranium ore, and access roads.

The proposed site will have a severely restricted, secure interior area where the ore will actually be processed. Here, workers will handle and grind the trucked-in ore, leach and thicken the yellowcake, and dispose of and monitor the tailings piles. This area will also house all operations involving uranium solvent extraction and precipitation and vanadium oxidation, all of which are highly technical and potentially risky processes. Buffering the high-security area will be secondary structures that include the administration building, labs, a house where workers will change their clothes before leaving the premises, and maintenance buildings such as a warehouse and a truck-repair shop.

The mill will handle about five hundred tons of local uranium and vanadium ore per day. The uranium will eventually be used for nuclear power production, probably in China, and various medical and industrial purposes, while the vanadium will harden steel so it can be used to manufacture batteries, alloys, and chemicals. Though supporters see the mill as a potential economic asset, its maximum estimated term of operation is forty years.[11]

Ore will arrive by truck from local mines. Once at the mill, the trucks will drive onto a scale, move to the ore pad to unload, pass a required radiation screening, and then leave the property to retrieve their next load. The uranium ore will be unloaded into the feed hopper and transferred via conveyor belt into the mill's main grinding and leaching building. There the ore will be dumped into a semiautogenous (SAG) mill, which uses steel balls to grind the ore to a fine dust that it mixes with water to create slurry. The slurry will be moved to storage tanks full of sulfuric acid for preleaching and then to a large prethickener tank. Leaching will take place in eight rubber-lined, agitator-filled steel tanks that will use toxic steam to heat the pulp that was pumped from the prethickener tanks. A combination of moisture and sulfuric acid will dissolve the uranium for final processing. From the leach tanks, the pulp will be sent to thickeners, about forty feet in diameter, which will separate liquid from solid. The waste pulp will be sent to the tailings ponds; and the second mixture, a uranium-heavy solution, will move into the extraction and precipitation processes, conducted in a separate building on the site. After filtration, that solution will be emptied into a uranium feed tank, constantly refreshed by hydrogen peroxide, that will precipitate the material into yellowcake (U_3O_8). The yellowcake will be dried, weighed, and packed into fifty-five-gallon drums for transport and further processing at offsite conversion facilities that enrich the material for energy production.[12]

"The Most Environmentally Friendly Uranium Mill in the World"

Crushing ore creates a significant amount of dust and increases workers' exposure to harmful radon progeny and other carcinogens.[13] But according to Energy Fuels and the Colorado Department of Public Health and Environment, Piñon Ridge Mill improves upon previous milling designs because the facility will use a SAG mill, a new technological advancement that helps encase uranium ore as it is pulverized, thereby reducing dust. Here and elsewhere, Energy Fuels has asserted, the mill will prioritize worker safety and health. In its proposal, the company emphasizes extensive safety measures, including numerous air monitors to track uranium dust.[14] Workers in especially dusty areas will wear special breathing-zone monitors, which will track the air they breathe at every moment. Those with especially risky roles will submit daily urine samples so that officials can monitor their exposure to inhaled and ingested uranium, and they will be required to wear gamma radiation badges to further track their exposure levels. In addition, personnel will follow regular equipment-cleaning and maintenance schedules, and the company says it will carefully measure and monitor hours of working-level exposure to uranium.

According to Energy Fuels, employees at the Piñon Ridge Mill will be environmental stewards in their treatment of tailings and radon levels. This is important publicity because even in Nucla and Naturita residents have many environmental concerns. In their answers to my questionnaire, residents worried about the mill's high water consumption in an arid region, changes in land-use patterns, radiation exposure, air and water pollution, chemical contamination, effects on wildlife and wetlands, increased truck traffic, noise, and general waste management.[15] Tailings and radon gas are particularly serious threats because milling produces two potent forms of radiation, radium-226 and thorium-230. Operations that release radon gas create ambient radiation that may affect workers and other populations near the mill. The corporation claims that it will continually monitor onsite radon levels to assure that nearby populations will be minimally affected. They also note that the mill's remote location will further ensure that radon gas concentrations and other pollutants will be diluted in valley communities.

Tailings piles can be linked to both air and water pollution, but in theory Piñon Ridge's tailings ponds will avoid many of the problems associated with older piles, such as those in Monticello, which were sometimes left uncovered. The Piñon Ridge site will include multiple ponds, where tailings will be stored as a slurry to avoid dust.[16] Each pond will be able to hold about 2.3 million tons of tailings and will have a multiple-liner system to avoid leaks into underground water aquifers, both during the mill's operation and after its closure.[17] To minimize radon gas emissions, each pond will have a cap that can withstand erosion over time. It is important to remember, however, that Energy Fuels will often be monitoring its own regulatory compliance.[18]

A Socioeconomic Savior?

Much of the publicity around the Piñon Ridge Mill has concentrated on the jobs the facility may create. But while supporters in Nucla and Naturita tie industry success to the robust local economies and thriving communities they recall from the 1950s, other area residents and regional activists focus on the uranium market's boom-bust volatility. Even if the mill does operate for its full forty-year span, it may not operate continuously or under Energy Fuels management or ownership. Due to lulls in global uranium markets and commodity prices, Energy Fuels' White Mesa Mill currently runs only about 60 percent of the time and even then must sometimes process alternate feed materials.[19]

Some influential reports have made optimistic projections about the mill's local economic effects. For instance, Montrose County's socioeconomic impact study concluded that an expansion in associated regional uranium mining and milling could create 516 to 649 jobs in the county. These new workers would, in turn, create 190 to 275 new households in Paradox Valley, increasing the number of residents by 32 to 46 percent. The report, cited by the Colorado Department of Public Health and Environment in its licensure decision, also suggests that the majority of new mill jobs will be high-paying (about 60,000 dollars per year, plus benefits) and that valley residents will be employed in most of these new positions. Similar reports, including those generated in consultation with Energy Fuels, assert that, unlike previous booms, contemporary uranium development will be more stable and secure for local residents. Yet the industry's continued instability and wavering market prices indicate that this one would probably be just as fleeting as the first two were. In fact, a socioeconomic report prepared by an economics professor at the University of Montana has major questions about the findings in Montrose County's report.[20] This study projects the total number of good-paying jobs at no more than 116, a figure that also includes jobs only indirectly related to Piñon Ridge.[21] It further suggests that the economic effects of mining and milling will be volatile and may have a negative impact on tourism, recreational outfitting, and second-home, amenity-based sectors.

The mill's highly mechanized design contributes to uncertainties about the facility's local socioeconomic effects. While supporters praise the mill's superior environmental and safety capabilities, their optimism rests on Piñon Ridge's complex mechanization and extensive use of technology. Thus, because computers, not people, will be doing many tasks, job projections may have been overestimated. In addition, those who do earn jobs at the facility may need skill sets that depend on advanced engineering, environmental, technological, and computer degrees, which local residents may not possess.

Although many mill supporters argue that employment and community participation in uranium extraction will strengthen economic security and wean the United States from its dependence on foreign oil, the yellowcake market itself creates other concerns. According to Energy Fuels, most of the yellowcake produced at the Piñon Ridge Mill will enter export markets and be shipped to China, whose fast-developing economy has created much of the demand for U.S. uranium. With most of the yellowcake going overseas, the United States may remain dependent on foreign oil, and uranium communities will remain embedded in unstable global markets.

Social Mobilization

In the late 1970s, accidents at Three Mile Island and Chernobyl triggered public hysteria about nuclear power's risks. Since then, nuclear facilities and storage areas have been notoriously difficult to site due primarily to public opposition. So when Nucla and Naturita residents mobilized in support of Piñon Ridge, they signaled an important shift in social activism even as they increased the legitimacy of both the project and Energy Fuels. Activists mobilizing support for the mill and industry renewal hold alternative notions of environmental justice; their site of acceptance centers on local autonomy over land use as a way to diminish persistent poverty. Activists believe that the mill symbolizes economic and community resilience within the structure of a familiar industry. Yet even though their point of view dominates in the valley's uranium communities, the area also has strong sites of resistance to renewed production. Opponents of the mill, who primarily live in unincorporated pockets of Paradox Valley and in communities such as Telluride that depend on tourism and amenity-based economies, believe that it symbolizes economic dependence, instability, and underaddressed environmental legacies and risks. As I discussed in chapter 1, their discrete sites of resistance coalesce to mobilize a Polanyian double movement: a fight for stronger socioenvironmental protections from free markets. Sites of acceptance, on the other hand, coalesce into triple movements, in which grassroots organizations privilege markets that they believe are part of their social fabric and trust those markets and involved corporations to self-regulate.

Sites of Acceptance

Sites of acceptance have arisen in the uranium communities that are closest to the proposed Piñon Ridge Mill site. The pool of activists is comprised largely of Nucla and Naturita residents, former Uravan residents who still live in the area, business owners and political leaders in Nucla and Naturita, and long-term, "fourth-generation" residents across the region who strongly identify with the industry. Their mobilization is most evident in their vocal support for the industry at public meetings and in public spaces. They post pro-mill signs in windows, businesses, and public buildings such as the Naturita visitor's center. The weekly *San Miguel Basin Forum* publishes ads and articles in favor of renewal;

and local institutions such as the Western Small Miners' Association, the Rimrocker Historical Society, and the Nucla-Naturita Chamber of Commerce also publicly support the facility and industry renewal more generally.

Unlike the situation in Monticello, local activism extends beyond ambivalence or general acceptance of extractive industries. These mobilized supporters are convinced that uranium industry renewal will breathe new life into their isolated and impoverished communities, regenerate local pride, and contribute to a patriotic energy future. Many supporters feel personally and historically connected to the industry. Rather than primarily linking it to social dislocation, boom-bust economies, or contested illness and environmental injustice legacies, they tie uranium production to pride of place. This attitude is particularly strong among former Uravan residents (see chapter 2). Uravan was destroyed because of extreme radioactive contamination, cancer clusters exist throughout Paradox Valley, and many local widows tell stories of miner and miller husbands who died of lung cancer.[22] But instead of focusing their activism on environmental health legacies or further remediation of radioactive sites, these former residents laud the memory of how an almost forgotten town contributed to atomic technologies and the rise of U.S. political power.

Activists have been working hard to build a museum, a campground, and a memorial on the site where Uravan once thrived. After years of negotiation with UMETCO Minerals, Dow Chemical, and Montrose County, the Rimrocker Historical Society (based in Naturita and composed of community historians and staunch industry supporters) has won the right to lease a parcel of land in the contaminated zone.[23] Soon Nucla and Naturita residents will work with society historians to construct a three-story replica of the Uravan Boarding House, which will serve as a museum commemorating Uravan, uranium workers, and "the Uravan Mineral Belt's rich mining heritage."[24] During one of my visits, members of the historical society took me on a tour of the old Uravan site. The woman leading the tour had grown up in the town and showed me where the schools, the swimming pool, the Little League field, and the neighborhoods had been. Her tears welled as she pointed out places that had held deep personal meaning. For her and many other former residents, a sense of lost community is palpable. Activism such as these efforts to commemorate Uravan fortifies and enriches sites of acceptance. For many supporters, challenges to the industry are the ultimate environmental injustice because they threaten community survival and local autonomy over community wealth: their land and its uranium.

Sites of Resistance

Sites of resistance to the Piñon Ridge Mill and uranium industry renewal have mobilized across the Uravan mineral belt and into other regions of Colorado. Although the Paradox Valley Sustainability Association has formed locally, most mobilization has occurred outside the valley, where it is anchored by the Sheep Mountain Alliance in Telluride and supported by dozens of organizations, including Uranium Watch, Western Colorado Congress, Colorado Citizens against Toxic Waste, and Grand Valley Peace and Justice. Sites of resistance extend beyond these regional groups into tourism and outfitting businesses such as Dvorak Expeditions; tourism-focused communities such as the town of Ophir; and national social movement associations such as the Helen Caldicott Foundation, the Endocrine Disruption Exchange, and the Toiyabe Chapter of the Sierra Club. In total, about sixty organizations, with a collective membership of nearly a million members, have been jointly communicating with the Colorado Department of Public Health and Environment during the mill's permitting process.

In January 2011, immediately following the department's decision to permit the mill, this coalition submitted a letter to their top administrators and regulators, predicting that "approval of this mill would likely lead to degradation of the environment, economy, and health of the region" and asking the department to deny Energy Fuels' license application.[25] In the following month, some of these groups challenged the corporation's permit in court, asserting that neither the company nor the department had created adequate space for open public commentary on the proposed mill. As I have discussed, regulations require that the department must hold a public hearing when considering such facilities so that citizens and organizations may call witnesses, cross-examine them, and receive formal answers given under oath. Instead, Energy Fuels and the department held multiple public *meetings*, where corporation officials typically gave twenty-minute talks describing the site, the mill, and details of the regulatory and permitting processes. Members of the public were then allowed to speak but were limited to three minutes each. According to attendees, the atmosphere was often restrictive, Energy Fuels officials and department administrators did not consistently respond to questions, and people who opposed the mill were discouraged from speaking. In November 2012, opponents' arguments were upheld. As the Sheep Mountain Alliance's director commented, "after years of trying to work with Colorado state regulators, who regularly ignored our concerns and still refuse to release key information, we are grateful for the opportunity to directly question the state about the Piñon Ridge Uranium Mill."[26] Nonetheless, in April 2013, despite new rounds of public hearings, the Colorado Department of Public Health and Environment reissued Energy Fuels' license to construct the mill; and the legal battles continue.

Class-Based Social Tensions

Sites of acceptance and resistance have mobilized along class-based lines that emphasize longstanding tensions among communities. Historically the most salient conflicts have arisen between the residents of Telluride and those of Nucla and Naturita, who are separated by sixty miles, several canyons and mountain ranges, and millions of dollars in resources. Those conflicts began as Telluride, a former mining town, shifted to a tourism-based economy in the late 1970s, drawing new businesses and new forms of wealth. As Telluride reinvented itself, it began requiring the services of not only contractors and construction workers but also low-wage laborers such as housekeepers, food servers, and resort workers. At the same time, Nucla and Naturita were descending into chronic recession as the uranium market busted; and by the 1980s Nucla had become became a bedroom community for the resort town. Residents made the windy and treacherous two-hour drive from their impoverished town to Telluride, where they worked in service jobs, struggling to make decent wages as they toiled alongside ostentatious displays of wealth.[27]

In 2000, as the town's wealth multiplied, citizens and other interested parties formed the Telluride Foundation as a charitable institution to improve the community well-being of not only Telluride but more isolated communities such as Nucla and Naturita. Thanks to generous grants and donations from founding members such as General Norman Schwarzkopf, the foundation initiated extensive community development programs.[28] For example, it created the Paradox Community Development Initiative, which offers low-interest loans and expert technical assistance to small businesses in the region.[29] The foundation has successfully used market-based solutions for community development issues: it is now the largest funder of business ventures on the Western Slope.[30] It has also focused on communities such as Nucla and Naturita in the economically depressed western end of the county, which it identifies as preferred locations for institutional programs and giving. In January 2006 it established the Building Community Resources Initiative, which specifically targets Nucla and Naturita as underserved communities in need of short-term assistance and long-term infrastructural and service supports. Describing those towns as "geographically isolated, . . . rural, underserved, and economically depressed," the foundation hopes to support community development efforts such as after-school programs, a skate park, and other youth-centered programs, given "the lack of government resources and paid staff for extra projects."[31] Though the foundation's website mentions that the need for these programs is structured by parents' lengthy commutes, it fails to disclose that most of those commuters work in Telluride's low-wage service sector.

Since the permitting of the Piñon Ridge Mill, the Telluride Foundation has actively opposed the facility, offering instead to fund remediation projects that would employ Nucla and Naturita residents while addressing hundreds of abandoned uranium mines in the region. However, many of those residents view such initiatives as insulting or misguided attempts to impose values on their communities. In other words, while foundation programs may be well intentioned, they often aggravate class-based tensions. Further, the extent to which the foundation's efforts actually aid these struggling communities remains to be seen. Through a program known as the Telluride Venture Accelerator, the foundation has offered recent awards to Hoggle Goggle, a Telluride-based ski and outdoor equipment startup; High Desert Farms, an organic and natural food operation in Dolores; Globa .li, a Denver-based online travel service; and Hyperlite Mountain Gear, an outdoor outfitter in Maine. New initiatives involving Nucla and Naturita are conspicuously absent from this list.

As grant programs to redistribute wealth collide with efforts to renew uranium production, tensions multiply among communities and activists. Divergent sites of acceptance and resistance instigate and perpetuate social tensions rather than align goals. Hampered by this difficult relationship and embedded in market-based, neoliberalized relations of production, activists on both sides have done little to revolutionize social or economic systems or to fundamentally shift social inequalities. In a structural sense, neither set of activists can do much to alter spatially unequal economic

PHOTO 6 Naturita's sparse Main Street reflects the town's struggling economy. (Photo by Stephanie A. Malin)

structures in which local workers provide raw materials—extracted from the ground, or sold as ski meccas and winter wonderlands—for consumption. They have little power or capacity to assuage structural violence related to natural resource dependence, with activists in sites of resistance motivated in part by their own tourism-based economies and desire for pristine environmental images. That said, those who are mobilizing sites of resistance are pursuing more transformative goals, including increased state protections, better access to healthcare systems, decreased dependence on unreliable natural resource markets, and the legitimizing of environmental health concerns. On the other hand, given their rurality and isolation, the lack of social safety nets, and persistent poverty, natural resource–dependent communities such as Nucla and Naturita are constrained to accept renewed uranium production.

Driving the Boom

In October 2011, several months after the Colorado Department of Public Health and Environment granted Energy Fuels Resources a permit for the Piñon Ridge Mill, the corporation orchestrated a friendly merger with Titan Uranium.[32] By making that deal, Energy Fuels became the owner of an estimated 31 million pounds of uranium reserves. In June 2012, the strategic acquisition of all of Denison Mines' U.S. assets gave Energy Fuels ownership of several key facilities, including the White Mesa Mill. Today the company oversees mines and development projects throughout the Colorado Plateau: the Whirlwind Mine and the Sunday Complex in Colorado, the La Sal Complex and the Daneros and Henry mines in Utah and Arizona, the Pinenut and Canyon mines in Arizona, the Sheep Mountain Mine in Wyoming, the San Rafael Mining District and the Energy Queen and Sage Plain exploration projects in Utah. Internationally, the corporation is also involved in copper exploration in the Canadian province of Newfoundland.

As Energy Fuels drives renewed uranium production on the plateau, it also exerts significant influence over the uranium communities closest to its operations—an influence that extends into the activism that is emerging in those spaces. The company's strong ties to Cold War–era production legacies have strengthened its reputation among supporters even as they have catalyzed distrust among activists who resist the industry's renewal. Although the corporation is not itself creating social activism related to the mill, it is certainly encouraging sites of acceptance. But by shaping opportunities for economic and community development in the Paradox Valley and establishing itself as a trusted institution, the company is also constraining the possibilities for open public discourse on issues related to renewed uranium production.

A New Company with Old Legacies

Company founder and former CEO George Glasier transformed Energy Fuels Resources into the impressive corporation it is today. By industry standards, the company is still a baby, which makes its leadership role in domestic uranium production even more unusual. Glasier did not start Energy Fuels until late 2005, and it did not go public on the Toronto Stock Exchange until 2006.[33] Yet despite the corporation's youth, both Glasier and the company behind Energy Fuels Resources have long ties to the U.S. uranium industry. Not only do these ties explain the current company's success, but they reveal its strong links to early uranium development and its legacies.

George Glasier provides a strong link between the industry's early incarnations and its current U.S. renewal. During the four decades he spent in the industry, Glasier accumulated extensive experience, wealth, and connections while also establishing himself as a trusted local in Nucla and Naturita. During our interview at his expansive ranch in Nucla, he told me that, after earning his law degree from University of Denver, he went to work for Bob Adams as a natural resource and mining lawyer. Glasier described Adams as a "pioneer in the uranium industry [who] . . . started clear back in the 1950s, owning one of the first uranium companies called Western Nuclear." In the early 1970s, as private companies were gaining access to uranium markets, Adams began a company called Energy Fuels Nuclear (linked only in name to today's Energy Fuels Resources) and hired young Glasier, just out of law school, to help him run it. Working as the company's lawyer, Glasier helped identify and acquire valuable uranium assets across the plateau.

By 1980, Energy Fuels Nuclear had completed construction on the White Mesa Mill. Subsequently, however, the mill was accused of complicity in several serious cases of environmental injustice, including assertions that its construction had disturbed sacred sites of the Ute Mountain tribe and allegations that the facility is still contaminating groundwater.[34] At the same time, uranium price fluctuations had affected mill operations, creating unpredictable operating schedules and unstable employment. Despite these controversies, however, by the mid-1980s Energy Fuels Nuclear was the biggest uranium-producing company in the United States, and Glasier had become a minority partner.

Bob Adams died in 1982, but Glasier and other investors kept Energy Fuels Nuclear in "the family," as he called it during our interview, for another decade. Glasier expanded his role, staying on as legal adviser while acquiring new responsibilities as public relations director and vice president of marketing. Then in 1991, the Adams family sold the company to a uranium commodities trader who, Glasier told me, conducted "a series of bad deals." This included borrowing about 2 billion dollars' worth of overstocked uranium from U.S. utility companies and selling it off at low prices. The new owner expected to replace the borrowed uranium with cheaper uranium sourced from Russia. But when uranium prices rose in the early 1990s and Russia refused to sell, both the owner and Energy Fuels Nuclear went bankrupt.

In the wake of one of the biggest bankruptcy declarations in Colorado's history, the company shut down all operations, including White Mesa Mill and every company mine. Hundreds of workers were laid off, and other uranium corporations began to acquire the company's substantial holdings on the plateau. International Uranium bought the White Mesa Mill for "almost nothing" before merging with Denison Mines in the late 1990s, and Denison continued to own and operate the mill until it was bought out by Energy Fuels Resources. In other words, Glasier's new company acquired a mill that had once been owned by his first employer.

Glasier had left Energy Fuels Nuclear before the bankruptcy occurred. Recognizing that the uranium market had bottomed out, he turned his attention to cattle ranching outside Nucla, where the locals—notorious for distinguishing between fourth-generation residents and newcomers—welcomed him. To earn that uniquely valuable sign of social acceptance, he spent years earning trust as a community fixture and an asset. Since buying his ranch, he has acted as a pro bono lawyer for residents and the town. He and his wife donate extensively to local schools, the chamber of commerce, and local and regional small businesses. He occasionally shares a beer in a town pub, and recently he was elected to the Montrose County hospital board. As a rancher, he has enhanced his credibility by means of a strong work ethic and a deep understanding of rural western life.

When Glasier founded Energy Fuels Resources, he also began acquiring area properties and amassing uranium reserves. Although its name was close to that of the bankrupt Energy Fuels Nuclear, he was certain that the similarity would not tarnish his new company's image: "Energy Fuels [Nuclear] had a great reputation, even though it went into bankruptcy. All of the contracts were satisfied. . . . The utilities were totally happy with Energy Fuels [Nuclear]. . . . We licensed the [White Mesa Mill] in Utah. We had a great reputation with the Utah regulators as well as the [Nuclear Regulatory Commission]. We were also the largest coal-mining company, called Colorado Energy Fuels. . . . I liked the name, and I liked the reputation. So . . . that's how I set up the [new] company from scratch in 2005."

In 2006, as uranium prices rose steadily, Glasier took Energy Fuels Resources public on the Toronto Stock Exchange. With that move, he was able raise about 60 million dollars, enough to acquire properties and begin the licensure process for constructing the Piñon Ridge Mill. Energy Fuels operates as a Colorado corporation, and Glasier believes that its focus on expanded uranium development and the construction of the Piñon Ridge Mill will benefit local communities. Throughout our interview, he emphasized the company's U.S. identity, despite its presence on both the New York and Toronto stock exchanges: "Energy Fuels Resources is a U.S. company. It owns all the assets. It's owned 100 percent by the Canadian company, [but] when people say it's a Canadian company, it's really a Colorado corporation. All of the assets are owned by the domestic corporation. It's just like General Motors has shareholders all over the world. So do we. . . . [Going public in Toronto] was just a way to raise money."

With offices in both Nucla and Kanab, Utah, Energy Fuels has worked to establish a significant local presence beyond Glasier's honorary status as a community member. During public meetings and informal gatherings with residents, company officials note that most members of their management team have lived in Colorado for many years.[35] Glasier knows that residents who support the mill link uranium with local pride, identify with the industry, and see the company as a reputable, trusted source of employment:

> They have a heritage of mining here, and that is what people, a lot of them, pointed out [at meetings]. Right now that is gone. . . . The ranching community is pretty tight, [but] the people who were in mining had to either leave or they are just barely holding on. They have to go into construction up in Telluride or whatever, waiting for mining to come back. But I think it will. . . . The Uravan picnic . . . has been dying, fewer and fewer people, [but] this year they had two hundred people show up because the community is coming back. The mining community. I see that, definitely more of a sense of community. This is a mining community.

Although he expressed similar optimism about the uranium industry's ability to enhance local economies and employment opportunities, he noted that economic benefits will depend on extensive industry expansion and unstable natural resource markets. He also added a caveat that he otherwise rarely mentions: significant job increases in the region will require the construction of a third mill, probably near Green River, Utah. Moreover, he told me that even though expanded uranium production suits the region, it will again make it vulnerable to boom-bust cycles:

> Before, in Utah and Colorado, there were six operating mills at one time. So this area has produced a lot of uranium in the past. People had good jobs, but of course, when the uranium industry shut down because of low prices, . . . the jobs went away. If the uranium comes back, which I think it will, if you could build another mill, then the jobs would come back. . . . It won't be like it was, but if we have three uranium mills, it will create . . . at least three hundred direct jobs from the mills and probably four hundred with the mining, the trucking, and the milling. Three hundred to four hundred direct jobs here.

PHOTO 7 A former Uravan resident looks across the highway to the place where her town once stood. Waste materials from Uravan's remediation—parts of old homes, buildings, and streets—are now stored under the white rock visible in the middle of this photograph. (Photo by Stephanie A. Malin)

Clearly, Glasier believes that renewed uranium production will bring economic growth to this spatially isolated area. He told me that he created Energy Fuels Resources to do exactly that: to bring life back to communities that have been faced with persistent poverty since the mining industries busted in the 1980s. "There is nothing else," he said. "All the construction in Telluride has come to a halt. So even if you were a carpenter making good money, you're not doing it now. The community is just dying, so the schools are going to die, and you won't even have medical facilities here." Invoking his honorary status as a local, he continued, "We like living here, being somewhat remote, but we want a store, we want a clinic, we want good schools." His conviction that renewed uranium development presents the best, if not the only, opportunity for uranium communities is persuasive.

Glasier retired as Energy Fuels' CEO in April 2010, shortly after county commissioners rezoned the Piñon Ridge Mill site. However, he remains a fixture in the industry, in the company, and in local imaginations. From his home in Nucla, he has undertaken several uranium-related consulting and business ventures with area residents, which has added yet another dimension to his community influence. Those partnerships include Varca Ventures, a mineral exploration firm comprised of local mining lobbyists and business leaders.[36] Yet he remains aware of unstable market systems and has hedged his bets in case Piñon Ridge and other such mills are never constructed. In 2012, he served as CEO and president of American Strategic Minerals, which has partnered with Ablation Technologies of Casper, Wyoming, to develop a new form of extraction and processing in which water and innovative mechanics separate uranium-bearing from uranium-free ore. If successful, this method could drastically reduce the demand for conventional facilities such as Piñon Ridge.

Recent downturns in uranium markets have brought Glasier back into close contact with Energy Fuels Resources and the Piñon Ridge Uranium Mill specifically as the corporation sells off risky or politicized assets in response to the market slowdown. In July 2014, CEO Steve Antony, whom Glaiser had handpicked to run the company before he left in 2010, announced that Energy Fuels was considering selling the Piñon Ridge Mill and its related permits to Baobab Asset Management. This development is significant. Glasier started Baobab after his departure from Energy Fuels, and a sale will firmly cement his central role in the fate of Piñon Ridge as well as the mill's ties to industry legacies. Although the sale has yet to be finalized and repercussions of the transfer remain to be seen, the development reinforces local sites of acceptance by enhancing Piñon Ridge's regional identity and legitimacy. As his new company gears up to take over the mill, Glasier has been invoking this connection, saying that the sale will return the mill to its local roots. Yet this shift in ownership, along with continuing delays in the mill's construction, highlight the instability of global uranium markets and the likelihood that those markets will bust again. Perhaps this will happen even before uranium communities have the chance to reap the rewards of a booming mill.

Meeting the Critics

In a neoliberal context that rewards private investment and free markets, Energy Fuels enjoys an enhanced degree of power due to its perceived ability to offer the social safety nets that the federal government once provided—steady jobs, benefits, and community development plans—to persistently poor residents in uranium communities. The company's top officials have used that influence to fortify sites of acceptance that had already mobilized to support uranium industry renewal and Piñon Ridge. As Peter, an upper-level administrator with the company who attended most public meetings on the licensure of Piñon Ridge, told me, "We just haven't had the need to really go in there and drum up support. That support was always there. . . . Nobody in their right mind is going to approve anything where the majority of local people are against you, . . . [but community members here] are pretty receptive."

With support already in place, Energy Fuels was able set the tone in public meetings and information sessions, where it provided ammunition for residents who wanted to speak in favor of the mill. For example, company officials linked the mill to new jobs and postwar community glory and worked to establish both the company and state regulators as fellow locals. These powerful themes, which ran through each gathering, controlled the parameters of debate and discussion and elevated the company's perceived role in area communities. So did its generosity: in 2010, Energy Fuels hosted several catered information sessions; between 2007 and 2011, it funded barbecues; and throughout those years, it sponsored question-and-answer sessions in conjunction with the Nucla-Naturita chamber of commerce. In addition, company representatives—usually Pete or George Glasier—attended each of the sixteen public meetings, where they sat prominently in the front row or onstage.

At each meeting hosted by the Colorado Department of Public Health and Environment, Energy Fuels officials had about twenty minutes to present their perspectives on the mill and uranium production, while all other individuals were limited to a strictly enforced three minutes each. Here and in their own information sessions, company representatives frequently mentioned the history of uranium activity in the region, referred to "long-time Colorado residents" on their main staff, and identified Energy Fuels as "a local company with strong ties to western Montrose County."[37] During one presentation, Pete assured the residents, "You know uranium better than anyone else, we can mill or mine it safer than anyone else, . . . and the mill will provide jobs. Of the eighty-five jobs we put in for at the mill, 80 percent of the workforce is expected to come from the local population. . . . Uranium is domestically abundant, it's clean, and we all know it."

In a context highly constrained by natural resource dependence, persistent poverty, and spatial isolation, Energy Fuels has been able to govern the space available for public debate, which in turn has reinforced tensions between haves and have-nots in the region. In fact, Energy Fuels has instigated those tensions. During public gatherings, administrators have firmly linked industry employment to local pride and culture while portraying residents' service-based work in Telluride as insulting and demeaning labor for the wealthy. Recalling his presentation at one of the company's public information sessions, Pete told me, "By bringing back mining and milling, we also bring back their culture. These people are comfortable with it. . . . I think it will be much better as far as families are concerned, being close-knit and making for a much better social situation. . . . It gets depressing when your options are so limited. You can drive to Telluride and clean toilets, but it's better to have something in your own backyard that is yours." *Cleaning the toilets of people in Telluride* has become a refrain in public meetings and information sessions. The phrase aggravates social dislocation; but by stirring class consciousness and spotlighting the angst of persistent poverty, it also motivates local activists to cultivate sites of acceptance for industry renewal, where local control over land and its resources becomes synonymous with environmental justice.

Prominent officials from the Colorado Department of Public Health and Environment have also noticed the effectiveness of using class-based tensions to fortify support for Piñon Ridge's construction. For instance, Will, who works in public relations at the department's Hazardous Materials and Radiation Management Division, told me,

> One of the other things at work here I've noticed are class divisions. In San Miguel County [home of Telluride] you have well-to-do, liberal people, . . . and in Montrose County [home of Nucla and Naturita] . . . you have a more conservative political outlook and working-class people in a depressed area. . . . People in the West End often have to drive over to Telluride to work . . . in low-end jobs, low-wage jobs. And the mill will provide higher-wage jobs, so there's kind of a class thing going on there, too. . . . So the way people think in the West End, I would suspect, . . . is . . . "Oh, okay, over in San Miguel County, they think it's okay for us to have jobs as long as it's six dollars an hour swabbing their toilets. But when it comes to a twenty- or forty-dollar-an-hour job at the mill, they don't want that."

In interviews, activists in sites of resistance have emphasized that they do not want to take jobs away from Nucla and Naturita residents, although

PHOTO 8 Telluride, once a mining town, has transformed itself into a ski haven and an expensive resort community. (Photo by Stephanie A. Malin)

they are well aware that Energy Fuels is deliberately aggravating class-based divisions to foster that perception. Maureen, the former mining director at Sheep Mountain Alliance, said,

> Energy Fuels has done their share of increasing the divide [between groups]. All along they have promoted, "Oh, now you guys won't have to go to Telluride anymore and clean condos." And so I think their strategy has been to divide and conquer. And they've done a pretty good job. So as far as [our] directly communicating with people in Nucla and Naturita, it's not very possible. It's not very feasible. . . . Not that there wasn't already a divide with the economic status difference, but it's been exaggerated. Energy Fuels has *totally* played that up.

Heather, an upper-level administrator at Sheep Mountain Alliance, concurs. After attending every public meeting and information session about the mill, she has come to believe that class-based arguments have deepened the resentment between groups that might once have been able to work together to develop other types of economic development, such as the funded remediation of abandoned mines. Yet she told me that such

PHOTO 9 The bustle of Telluride's Main Street is strikingly different from Nucla's and Naturita's quiet, empty town centers. (Photo by Stephanie A. Malin)

divisions are exaggerated, explaining that her own class position is not so different from that of many people in Nucla and Naturita:

> They do want jobs. I mean, these people have to drive a long distance to come and find work in Telluride. And they are the same kind of jobs I had when I first started here. I cleaned toilets for ten years! . . . So I understand where that comes from. . . . But I work four jobs. I'm a single mom, I have a daughter, and I've spent the time to make factual arguments about the mill. . . . The next person who calls me a trust funder at one of these meetings—I take offense at that! I wish I was a f—ing trust funder! . . . But I don't go in there with that because I understand their frustration and I feel their pain. . . . We need to get those people jobs.

In short, while highlighting class-based inequities does strengthen support for Energy Fuels' cause, it also increases regional social tensions that inhibit collaboration and communication among communities. Instead, activists must use courts and lawyers to relay their concerns about local land use, energy development, and definitions of environmental justice. In the meantime, activists in sites of resistance refer to activists in sites of acceptance as "those people." There is no personal connection.

PHOTO 10 Nucla's local economy has suffered since the uranium industry's last bust in the early 1980s. (Photo by Stephanie A. Malin)

Constrained Space for Public Discourse

Energy Fuels and the Colorado Department of Public Health and Environment promised that each stage of the Piñon Ridge Mill's permitting process would be transparent. But in public gatherings, Energy Fuels has capitalized on class tensions, an approach that state officials have failed to address or mitigate. In addition, before the court ordered them to do so, neither institution created public spaces in which residents could ask technical or substantive questions and receive answers from the company or the state. The blatancy of the situation led the Denver District Court to set aside the mill's license and call for a formal public hearing, though that license has now been reinstated.

Many activists in sites of resistance told me they were not able to share either their concerns or their notions of environmental justice during public meetings and information sessions. Heather, who has been a vocal leader against renewed uranium production, said that some mill supporters have bullied her at meetings, and those reactions have escalated over time. Yet even when they had direct knowledge of those tactics, Energy Fuels officials did not interfere. In an interview, Heather described one particularly threatening encounter.

> After the last [public meeting], I was basically corralled by four huge SUVs as the Energy Fuels guys walked by and sort of laughed at me and didn't offer to help or anything. So I had to get out of [my] car as people were staring me down and move rocks and huge boulders to be able to drive out of the way. I was watching in my rearview mirror that whole way home thinking, "F—, . . . what are my outs here?" I was watching the road in front of me to see where I could drive to that would be safe, and I gripped the wheel the whole time. And the last meeting was the worst. . . . I got up and basically said this has not been a public process and you [state] guys are advocates for [Energy Fuels]. . . . Then four or five people [got] up after me and [said], "That stupid trust funder from Telluride doesn't know what she's talking about."

According to Heather, the department does little to ensure that both sides can voice their concerns: "The state [of Colorado], it's their responsibility to run . . . public meetings where you are heard equally on both sides and it's a fair hearing. They've done nothing to encourage that. In fact, they've snickered a couple times when people make jokes about [Telluride residents]." I heard similar remarks throughout my interviews with activists who were opposing the Piñon Ridge facility. Some recalled meetings in which intimidating groups of police officers were lining the back of the room. Others mentioned beer-heightened tensions during public barbecues hosted by Energy Fuels. The Nuclear Regulatory Commission and the judicial system tried to remedy these systemic issues when they mandated the November 2012 public hearing, but that change came about only after activists in sites of resistance had formally challenged the department's permitting process in court.

Because activists have not been able to debate substantive issues in public forums, tensions have spilled outside public meetings and courtrooms, reinforcing the rifts between uranium and tourism- and amenity-based communities. As in Heather's case, some of those tensions have erupted in scary but fleeting interactions directly after meetings. Other hostilities have escalated over time. Dawn, a Paradox Valley resident, told me that she received numerous phone calls and death threats and had several frightening encounters at the local market after she publicly announced her opposition to the mill. For a time she refused to talk to me because of trauma related to these altercations and tensions, which have triggered severe anxiety and even led to a hospital stay.

Clint, a founding member of the Paradox Valley Sustainability Association, attended several open-forum meetings hosted by community members who were working to create space for public debate about the mill. All of these meetings were uncomfortable. He recalled that even in neutral settings, social tensions stymied efforts to create open space for discourse. The meeting that he hosted was a flop. During others, intimidating behaviors extinguished any hope of democratic discussion. He told me, "[One public official] had a meeting at the community center in Paradox, where he brought in his friends because he's on the West End sheriff's posse. And he brought in his friends from the Montrose County sheriff's office to stand at the door. And the first thing he told the crowd was that anybody who spoke out against the mill proposal would be escorted out of the building by the Montrose County sheriffs. . . . He's an amazing person. But seeing him react in such an extreme way was disturbing."

Clint attributes the intensity and duration of social tensions to the ways in which Energy Fuels has handled its role.

> This is the crux of the problem out here socially. The people who have high financial interests . . . are using divide-and-conquer tactics. They use fear and

> intimidation. . . . Energy Fuels, they want to make a profit, there's Montrose County who wants to make a profit, and [the state] is interested in creating an industry and keeping their jobs. . . . So you've got three well funded, politically powerful groups using their influence to create division in the community. . . . The influence creating conflict in this issue is a hundred time more powerful than the interest to create a working relationship between these two groups.

Residents of Paradox Valley must cope with a daily atmosphere of conflict. Jeanne, a wildlife biologist who lives in the valley and is a member of the Paradox Valley Sustainability Association, described the situation:

> I find that most people are afraid to speak out against [the mill]. And maybe that's because the group in favor, . . . there is a longer connection between all of them. They are kind of the bullies of the valley, in my opinion, in being very, very vocal about being in favor of the mill. . . . They'll say if you don't want it, you should leave. . . . I think most of the people that are opposed to the mill are intimidated to speak out because they don't want to be ostracized by this . . . core community that's so vocal—and *mean* about it! I mean, there have been people that have threatened . . . , "If [the mill] doesn't happen, there are people that are going to die because of it." There have actually been threats.

Renewed Production and Social Activism

For activists, the Piñon Ridge Mill represents the risks and rewards of renewed uranium production on the Colorado Plateau. Touted by Energy Fuels and even the state of Colorado as an environmentally friendly and technologically advanced facility, the mill is a symbol, among supporting activists, of community development, good opportunity for corporate self-regulation, and familiarity. In contrast, activists in sites of resistance see it as a symbol of economic dependence on volatile natural resource markets, instability, and risk of environmental injustice. In this context, Energy Fuels Resources (and if the mill is sold to Baobab, perhaps George Glasier) is capitalizing on its power and influence in Paradox Valley. As the driver of renewed uranium production, the company offers a potential solution to residents' social dislocation. At the same time, it exacerbates class-based social tensions and instigates venomous reactions to activists who resist the mill.

Hegemonic neoliberalism enacts structural violence in two different ways here. First, Energy Fuels has masterfully used its local identity and frequent public interactions to establish itself as the institution that will offer desperate residents full-time jobs, healthcare, and other social safety nets. Its explanations appeal to market-based logic and neoliberal discourse, both of which have become normalized in the United States. By encouraging residents to weigh their individual needs for employment against collective concerns about environmental degradation and health outcomes, the company cultivates sites of acceptance in which less transformative types of activism mobilize.

Second, Energy Fuels downplays the region's historical vulnerability to chaotic boom-bust cycles, economic instability, and debilitating social dislocations. Not only does the company instigate class-based tensions with wealthier communities, but it also reminds locals that the industry and its unstable markets have become part of their identity. At the same time, by presenting the Piñon Ridge facility as a source of stable, environmentally safe jobs, Energy Fuels seems to promise that debilitating natural resource dependence will not be as severe this time around, that ultimately the industry will pull Nucla and Naturita out of persistent poverty. In other words, the company encourages supporters to ignore historical patterns of economic desperation and to privilege market-based notions of environmental justice and activism.

5

"Just Hanging on by a Thread"

Isolation, Poverty, and Social Dislocation

The Nucla-Naturita area is like a sponge that's extremely dried out, and all we're asking for is a bit of moisture . . . to bring us back to some semblance of normalcy because you've seen what our downtown areas look like, and it's not a normal community.
—Philip, Naturita resident and mill supporter

I don't know if the jobs and benefits to [Nucla and Naturita] that the community imagines are going to happen. . . . I don't think jobs that pay 75,000 dollars a year descending out of the sky is going to happen, . . . and I'm not sure those would go to the people who really need the jobs, due to their qualifications. . . . It's snake-oil salesmanship to me.
—Don V., Telluride town councilman and mill opponent

The Colorado Plateau is defined by spatial isolation, spare landscape, and a sparse population.[1] It comprises about 140,000 square miles of awe-inspiring geologic anomalies such as waterpocket folds and slot canyons, some of the oldest exposed rock in North America, and more national parks and monuments than any other region of the United States.[2] Yet even as increasing numbers of tourists, retirees, and second-home owners make their way to area ski slopes, canyons, and parks, the core population density remains at fewer than five people per square mile.[3] A handful of heavily populated, urbanized areas—St. George, Utah; Grand Junction, Colorado; Flagstaff, Arizona—serve as commercial, medical, and entertainment hubs for the remainder of the residents scattered in rural communities across the plateau.

Spatial isolation defines daily life in the region's far-flung communities, limiting access to healthcare, economic opportunities, Internet connections, restaurants, media, and other amenities. At the same time, residents revel in the plateau's unparalleled beauty, its recreational opportunities, and its solitude. The place feels like a world apart: a driver may travel on a state highway and never see another car, may visit a national park and never see another individual. Montrose County, Colorado (home of Nucla and Naturita), has a population density of eighteen people per square mile; San Juan County, Utah (home of Monticello), has a density of two people per square mile.[4] In contrast, the state of New Jersey, which is roughly the size of San Juan County, has a population density of 1,185 residents per square mile, people who are connected via a vast web of interstates, public transport systems, and information networks to other high-density, high-powered population centers such as New York City, Washington, D.C., and Boston.[5] The residents of uranium communities, however, are often hundreds of miles away from urbanized areas with no options for rail, air, or other means of transportation to connect them to these hubs. Nucla and Naturita are more than two hundred miles away from Colorado Springs, the closest town with a population higher than 50,000.[6] For Monticello residents, the closest such town is Provo, two hundred miles to the north.

When natural or technological disasters strike already vulnerable rural pockets, as Hurricane Katrina did in the Southeast and as long-term nuclear contamination has done in the West, all people are vulnerable. But even in times of economic growth, spatial isolation restricts residents' access to basic services and social safety nets.[7] We have long known that systems of inequality structured through demographic variables such as race and poverty can multiply vulnerability, and legal scholar Debra Bassett suggests that geographic isolation is a crucial third variable, "caus[ing] the rural poor to be forgotten, hidden, and indeed repressed from view and memory, . . . not just powerless but genuinely forgotten to the point of invisibility."[8] Legal and political institutions tend to exacerbate the problem by facilitating arrangements that either ignore or underestimate rural poverty rates. Or they equate its dynamics with urban poverty and thus fail to address the unique problems of the rural poor.

In uranium communities, these components of disadvantage are complicated by natural resource dependence and the "slow-motion technological disasters" related to uranium, which deepen structural disadvantages and social dislocation.[9] Environmental sociologist Valerie Kuletz, who has tracked the environmental injustices of the early atomic age, calls these areas "invisible nuclear landscapes" where the native populations have been perniciously and structurally disadvantaged.[10] Resource geographer Arn Keeling, in his study of Uranium City in Saskatchewan, notes a link between social dislocation and the rapid boom-bust cycle of uranium development.[11] My own research reinforces these findings: uranium community activists can feel invisible as they struggle to develop their communities and interact with extralocal institutions. As a result, residents lose trust in federal institutions and scientists and begin to see their isolated communities as national sacrifice zones.[12]

Rural Dislocation and Poverty

Persistent poverty's invisibility in rural spaces can magnify the effects of isolation because poverty shapes and limits these communities whether or not it is fully captured statistically, a situation that exacerbates concerns about the validity of U.S. poverty measures in general.[13] To illustrate: the U.S. Census Bureau's American Community Survey, which provides the most definitive town-level data for American populations, estimates that Colorado has a statewide poverty rate of 12.5 percent. For Nucla, the survey records an above-average rate of almost 17 percent, with an associated margin of error of 10 percent in either direction.[14] Likewise, Naturita has an above-average poverty rate of 18 percent, with an associated 11-percent margin of error.[15] Even Monticello's less staggering poverty rate (10.5 percent) has an associated 6.4-percent margin of error.[16] In contrast, crowd-sourced data indicate that Nucla's poverty rate is about 30 percent, Naturita's about 20 percent.[17] My interviews overwhelmingly confirmed this. Wide margins of error and large discrepancies in these statistics underplay the persistence of poverty that is very real to Nucla and Naturita residents.

In addition to creating margin-of-error inaccuracies, census data on poverty often focus on counties rather than individual communities (see map 4). For instance, while the U.S. Department of Agriculture's Economic Research Service labels most Colorado Plateau counties as persistently impoverished (among them, Utah's San Juan County), it does not include Colorado's Montrose County in that ranking.[18] This further obfuscates the extent and impact of poverty in Nucla, Naturita, and other Montrose uranium communities.

Overall, such measurement problems highlight two issues. First, social scientists know little about the real rates of poverty or their persistence in spatially isolated uranium communities, and that ignorance continues to distance these places from appropriate policy resolutions. Second, despite apparent inaccuracies of U.S. census data, when combined with local narratives on poverty, they indicate that rural poverty does in fact persist at high rates in these communities, leading to chronic material deprivation for residents. Thus, material conditions such as spatial isolation and persistent poverty induce structural violence, where access to economic, political, and social resources is so severely constrained that uranium communities can neither sustain robust economies and populations over time nor exercise robust control over how to develop them.

Powerlessness and Divergent Activism

Persistent poverty, spatial isolation, and natural resource dependence have been consistent material conditions, but they have affected plateau communities in various ways. As residents have worked to address the possibility of uranium industry renewal, they have branched into divergent forms of activism reflecting their unique notions of environmental justice, economic need, and community identity.

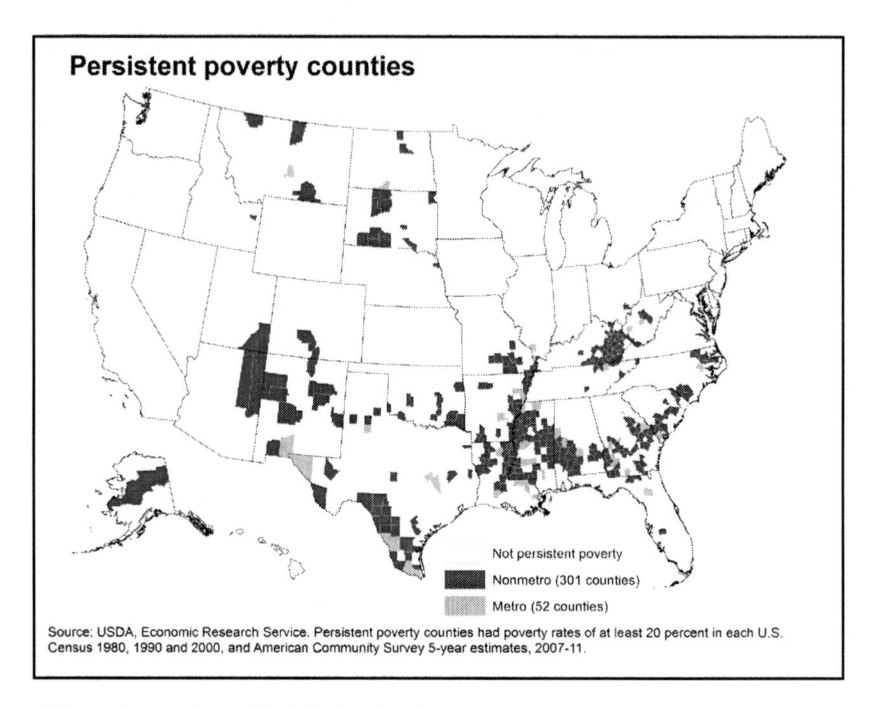

MAP 4 U.S. counties with high levels of persistent poverty.
Source: USDA Economic Research Service.

Material Conditions Shaping Sites of Acceptance

Telluride has used its spatial isolation as a springboard for transforming its natural resource dependence from mining into tourism and recreation. Today, wealthy tourists spend their vacations and their money in the town, thus shielding it from persistent poverty. Yet uranium communities such as Nucla and Naturita have been unable to parlay their dependence into a multimillion-dollar ski industry, partly because of aesthetic differences and partly because they have overadapted to extractive industries.[19] Every resident and community leader whom I interviewed in Nucla and Naturita described the psychosocial stress they felt, along with a chronic sense of economic vulnerability, as recession and poverty persisted and the uranium bust dragged into yet another decade. At the same time, their overadaptation to the industry meant that most were vocal proponents of renewal.

At a January 2010 public meeting hosted by the Colorado Department of Public Health and Environment, one Nucla resident said, "We have a

PHOTO 11 Chronic poverty in Naturita is reflected in substandard and temporary housing. (Photo by Stephanie A. Malin)

trucking industry to support [the mill]. We have the heavy equipment industry to support it. We're ready to step into the future and make it a good one. We need to be able to grow, but not at the expense of our miners and millers. . . . We have everything we need to make this successful, we're ready for it, and this area deserves it. We've been affected by the economy for so many years. Please say yes to the mill." Another said, "I used to work for Union Carbide. I was there until the mill closed in '84. Our mill closed, not because we didn't meet the state regulations; our mill was closed because of outside uranium that was brought in. . . . [This] affects the lives of those of us here tonight. We need jobs. When you get right down to the nitty-gritty, we need jobs."

A sense of social dislocation emerged prominently during my interviews. Lois, a community volunteer and lifelong resident of the area, described the destabilizing effects of persistent poverty, which her neighbors and her family normalized as an element of daily life: "As long as mining was going just a little, it wasn't too bad. Our trucking was going, and of course our power plant with the coal was going. We were never overly rich, but we weren't completely down either. We were just kind of hanging on. . . . At least we could do that while there was something going. But now there's nothing. So we're all just hanging on by a thread." I heard similar remarks from community leaders and entrepreneurs. Abigail, president of Nucla and Naturita's chamber of commerce and a local business owner, observed, "People are not spending money like they were. They are being very careful because they're not sure what's coming next. But having a mill here would be an enormous thing. . . . It's going to boost the economy all the way through." Like Luke, a small-operations miner who also leads a regional mining lobbyist group, most residents accept the industry's bust periods because it may, at least temporarily, assuage the community's economic struggles. Natural resource dependence becomes normal. Said Luke, "The industry gets all fired up; then the market completely tanks and crashes. You've got this tumor that's building, then all of a sudden it completely dissolves away, and the area has to contend with it. . . . That's something our kids have to contend with, too, and many cannot stay. That's just the nature of areas like Nucla and Naturita. . . . They are built on extractive industries."

Support for industry renewal persists in Paradox Valley in part because people believe that the chaos of the market cycle is normal. In a household survey I conducted in the four communities closest to the proposed Piñon Ridge Mill site, 77 percent of respondents expressed *strong* support for construction, while 87 percent of people expressed at least *some* support for it. When I asked them to identify the three primary benefits related to the mill's presence, 94 percent of respondents listed "employment" as the primary benefit, and 52 percent identified "economic and community renewal" as additional benefits.[20] One respondent from Nucla wrote, "The mill will create jobs and lots of them. It will bring new and younger people to our area. This will spur support of businesses and grow our communities. Our schools will benefit with more students and younger teachers to boost community pride." Another respondent commented, "We need to support this . . . industry for today and the future if we want to go back to a working, thriving, smiling community."

Many residents of uranium communities have spent their entire lives in these isolated and impoverished towns. Their fathers often worked in the uranium industry; as children, most of them played on tailings piles. Old-timers love to regale young residents with stories of uranium prospecting during World War II and the Cold War. They love their communities because these places feel like home—which is why, for many, simply leaving is not an option they can imagine. For better or worse, uranium lies at the heart of these towns; and each time I stand in the middle

PHOTO 12 As Naturita loses population, locally owned businesses struggle. (Photo by Stephanie A. Malin)

of Naturita's dilapidated Main Street, I can understand why townspeople worry about their community's ability to thrive, or even survive. Such concerns permeated both the comments at public meetings and the long stories I heard during interviews. At a January 2010 public meeting hosted by the Colorado Department of Public Health and Environment, a Naturita resident displayed a series of maps depicting the demographic effects of chronic recession: "If you look at the historical population growth chart . . . provided by the Department of Local Affairs, you will see that over the past fifty years, while Montrose County has increased by 220 percent, the town of Nucla has decreased by 17 percent and Naturita [by] 30 percent. The absence of industry and primary jobs truly has a direct effect on a community's ability to thrive." Philip, a thirty-year resident of Naturita, concurs:

> The Nucla-Naturita area is like a sponge that's extremely dried out, and all we're asking for is a bit of moisture . . . to bring us back to some semblance of normalcy because you've seen what our downtown areas look like, and it's not a normal community. . . . All we're asking is to bring some normalcy to our community, and in thirty years, tourism hasn't done it; . . . the coal mine and the power plant haven't done it. . . . This mill will be a good thing for the community just to bring us back to some kind of normal point where we can call ourselves a community: we can have a school that we can be proud of, our kids

> will have a place they can be proud of because right now we're struggling with that.... I believe the mill will create 1,000 jobs.

As I discussed in chapter 4, George Glasier and Energy Fuels have worked hard to reinforce this perception. Glasier told me, "The mill is going to create enough jobs for the people here so that the community will boom again.... It will be like 1979 again. You know, the high school graduated seventy-nine students in 1979. Last year, I think they graduated twelve or thirteen. You can see what's happened.... The community is just dying. So the schools are going to die." Most of the locals seem to believe him. Tanya, Naturita's mayor, said, "I trust that George and Energy Fuels are going to have locals work for them.... George said this will bring another boom to the area . . . and just from watching George, you can tell he always makes sure he's always doing everything right. He's been the town lawyer for fifteen years, and he always just does things right and efficiently." Ben, a Nucla resident and a school board member, agreed: "Even though George's involvement in the mill process will be curtailed with his retirement, . . . he will still see that jobs go to locals." As I mentioned in chapter 4, one of George's new companies may buy the Piñon Ridge Mill's permit and associated assets, meaning that these perceptions will probably only intensify.

Rebuilding educational and employment opportunities for young people is a key concern. As Pete of Energy Fuels noted, "The school is particularly depressing if you look at the numbers of people graduating. So many kids end up going into the military because they really have no other options, or are migrating out." During interviews, many Nucla and Naturita residents told me that renewed uranium development—and especially a new state-of-the-art mill—could decrease the exodus of youth from the region. With few good jobs available locally, and many of those dependent on unstable natural resource markets, younger residents often do not return home after college.[21] Those who do not attend college often see military service as their best option, a choice that may permanently separate families and thus further strain the social fabric of the communities. During a February 2010 public meeting, one resident observed, "If you go to the malt shop and look on the wall, there's a bunch of pictures of military boys and girls.... And there's graduation pictures and little yellow ribbons around them or they've written where they are and, you know, we miss you.... And it's pretty touching to see that the same options for young kids

PHOTO 13 Due to declining enrollments, a shrinking tax base, and financial pressures to consolidate operations, Nucla and Naturita have had to sell valuable assets, including the Nucla school building pictured here. (Photo by Stephanie A. Malin)

in Naturita isn't the same as someone from the big city. I think we can do better than that. And I would like to see those kids have a better chance."

Reinforcing all of these material issues is a strong sense of confidence and patriotism. Residents believe they understand uranium better than most Americans do. As Don Colcord, the only pharmacist in a hundred-mile radius, said, "I think it's more important than anything else that we have been around it, we're not afraid of it, we don't have any preconceived notions that it's dangerous. There are very few places . . . you are ever going to find that would accept it as willingly as we would."[22] In his view, the area has a unique ability to help the United States address climate change, which he elevates above even employment incentives: "The country needs us as much or more than we need the jobs. . . . We're saying that we want this, we're willing to do this because we think nuclear power is the best way to . . . solve the country's energy problems." Tanya agreed: "Nuclear power is vital for our country, . . . and our country doesn't need to be buying these supplies from Russia and other places. We should be mining, milling, and stockpiling to take care of ourselves, . . . and we are the community to do it . . . because we know uranium here." An Energy Fuels employee who is also a Nucla resident explained, "We can go underground, we can mine and mill uranium, and we can produce a commodity for this nation . . . that is desperately needed. And we can do it safer than anybody else in the world." Some residents continue to believe that they are immune to any of uranium's debilitating health effects. Roxanne, Bedrock's postmaster and an avid mill supporter, commented, "I know a family that lived in Paradox Valley, and when their young son bit into yellowcake some years back, they took him to the ER and thought he was going to die. To this day, he's one of the most healthy, handsome men I know."

Nearly every mobilized uranium industry supporter I interviewed referred to a study commissioned by Union Carbide, which found that Uravan residents were no more likely than any other Coloradoans to die from uranium-related cancers.[23] Don told me, "That study showed that people in Uravan were actually healthier than those in the rest of America. I grew up there, so I know. . . . We were running around outside, being healthy. This study confirmed . . . that people living around uranium have no ill-health effects. And I tell my customers about that."

Material Conditions Shaping Sites of Resistance

Though most industry opponents also live in spatially isolated communities, they tend to live in towns where nonextractive natural resource economies such as skiing and tourism now thrive, or they are trying to initiate alternative economic activities such as sustainable agriculture. Importantly, most anti-mill activists are securely middle or upper class; and in part because their economic security requires a pristine landscape, their notions of environmental justice are strongly aligned with the distribution of environmental risk. Yet they also worry about continuing economic dislocation in uranium communities. At a January 2010 public meeting a San Miguel County resident attending a January 2010 public meeting argued, "These are not stable jobs. These people deserve better. And to build an economy again on uranium and to pin all these people's hopes and lives and their families on uranium, and watch it collapse next time we see a fluctuation in the price of uranium, I don't think that is appropriate for these people or this area." At the same meeting, a member of the organization Grand Valley Peace and Justice noted that "a socioeconomic report from the Sedora Institute, given

PHOTO 14 Naturita, Colorado, still strongly identifies with uranium. (Photo by Stephanie A. Malin)

to county commissioners, . . . talks about measurable negative impact to the region due to the boom and bust nature of any mining industry, including uranium."[24] At a February 2010 public meeting, an Ophir resident also expressed skepticism and anticipated the recent drop in uranium prices: "Let's assume the mill is built. . . . The worst situation is they operate and the employees feel secure, [and] then there's a drop in yellowcake prices and demand. Once again, we face a cleanup situation and a closed mill. Once again, the winds blow in the spring. The Piñon Ridge Mill does not appear to be a solution for employment on the West End, nor an overall attractive long-term development plan for citizens of Colorado."

Various people I interviewed across the region expressed feelings of economic vulnerability or powerlessness in relation to uranium markets and their historical instability. Clint, a local resident and co-founder of the Paradox Valley Sustainability Association, suggested that the Piñon Ridge Mill might *cause* social dislocation: "Why are Nucla and Naturita so distraught and poor and so devastated? The only reason I can see is because of uranium, and the idea that uranium is going to come in and make everything fine seems to me extremely questionable." Don V., who serves on the Telluride town council, expressed similar concerns: "I don't know if the jobs and benefits to [Nucla and Naturita] that the community imagines are going to happen. . . . I don't think jobs that pay 75,000 dollars a year descending out of the sky is going to happen, . . . and I'm not sure those would go to the people who really need the jobs, due to their qualifications. . . . It's snake-oil salesmanship to me."

Heather, current president of the Sheep Mountain Alliance, echoed these sentiments. She is acutely concerned about Energy Fuels' ability to hire as many local residents as it has promised in public meetings and socioeconomic reports: "They'll hire a couple [of] locals and the local construction company for sure. But they just shut down the Cotter Mill and those sixty or seventy people have direct experience.[25] [So] they are going to be bringing in a transient population, which will create huge social burdens on the communities. So the idea of eighty-five jobs is absurd." In partnership with the Telluride Foundation, the alliance believes that remediating abandoned uranium mines would create more long-term employment without subjecting people to uranium market volatility. Heather explained, "We're not anti-jobs, . . . but let's pursue these other options as well. . . . I understand [residents'] frustrations, and I feel their pain, . . . which is why we need to get these people jobs. We need to bring in different industries."

Opponents also fear that expanded uranium production will disrupt agricultural pursuits or destabilize tourism and recreational industries. Since the zoning of Piñon Ridge, some agricultural property values in Paradox Valley have already dropped, creating economic vulnerability among local farmers. Dawn, a sustainable seed farmer in the valley, said she will be compelled to move if the mill is actually built because it will damage the reputation of her seeds: "I love what I do and produce on this land, and it unnerves me to think of not being able to do this. But if this mill comes in, I will have to leave." But leaving may be difficult. As Jeanne, a local wildlife biologist and sustainable farmer in the valley, commented, "our personal experience with trying to sell our family farm has not been positive. I'm . . . moving to Oregon, but we've still been unable to sell our farm, which I love dearly, because of the rumor of the uranium mill." Telluride's ski-based economy depends on the region's natural beauty, and the presence of a uranium mill—even one sixty miles to the west—threatens the community's wilderness image.[26] Alliance president Heather, a Telluride resident, told me, "If the mill comes in, I may have to leave, which makes me very sad. I have a daughter, too. But it just makes me too nervous. I've heard from several other folks around here that they will do the same if the mill comes in, mostly because they work for the ski resorts and expect to lose their jobs if the mill is approved."

Activists in sites of resistance repeatedly shared their concerns about air quality, water scarcity, radioactive dust storms, and the health effects of production renewal. During a January 2010 public meeting, one attendee commented, "We've just gone through over a month of [air] inversions in Paradox Valley. . . . Over time, there will be concentration of uranium in that air when it is calm and sitting down in the valley. . . . Do you plan on sacrificing the population of Paradox Valley because it's small enough and that's an acceptable loss?" Dawn asked me, "Where are the air models that take into consideration the fact that . . . last winter we had almost two months of inversion through the valley, and we didn't have air leaving the valley? . . . What kind of impact is that going to have on local residents' health?" An attendee at a February 2010 public meeting questioned the practice of privileging uranium markets and employment above environmental sustainability: "Is the prospect of adding ten or twenty new jobs worth the very real risk of damaging the water we all drink and the air we breathe? We need to remember that more people are affected than just those in the Nucla and Naturita area. We're all in this together and should

PHOTO 15 While a working class powers Telluride's tourism-based industry, many of those individuals can no longer afford to live within the city limits of such a wealthy community. (Photo by Stephanie A. Malin)

support the efforts of the Paradox Valley Sustainability folks to find other jobs that don't risk our health and our environment."

Some opponents are themselves dealing with health issues related to uranium production. At a January 2010 public meeting a Grand Junction resident said, "I have cancer, as do many of my classmates and people I grew up with. . . . I know you can't keep uranium out of the water, out of the air, or out of things that we eat. I'm asking [the state] to consider this when you make your decision." Ryan, a former uranium miller and lifelong Paradox Valley resident, told me about an accident that had happened to him:

> Back in 1967, I fell into a uranium tank at Atlas Mills over in Moab. It's a big vat—twelve feet by twelve feet . . . I was working graveyard, and swing shift had removed the catwalks and hadn't roped it off and the middle light bulb was out. . . . So I was carrying out a light bulb and all of a sudden fell twelve feet deep into that nitric acid, sulfuric acid, caustic soda, and yellowcake. . . . If I hadn't been a good swimmer, I would have stayed in there. . . . So that's my

> biggest concern is safety . . . when you are working twelve or sixteen hours. . . . They can practice safety all they want, but there are going to be accidents that could impact surrounding communities. I have suffered from rashes ever since, even in my eyes and throat.

According to uranium expert Doug Brugge, perceived environmental and health effects may be just as psychosocially damaging and dislocating as actual exposures are.[27] Recalling his research on uranium exposure on the Navajo Nation, he said, "It doesn't matter, to a certain extent, if there aren't any dust emissions if people think that there are. I brought that up at a meeting with Montrose County commissioners, and the county seemed all over it, . . . but then permitted the rezoning anyway." Brugge further observed that the existence of psychosocial dislocations is supported by many peer-reviewed studies that provide strong evidence for uranium's health effects, particularly on kidney functioning and breast cancer.[28]

In open-ended comments on the survey instrument I distributed throughout Paradox Valley communities, environmental and health-related dislocation emerged as significant community-wide concerns. When I asked respondents to identify risks associated with the Piñon Ridge Mill, 28 percent named "general safety concerns," 39 percent identified "environmental contamination," and 10 percent identified "health risks." One respondent believed that health concerns were the biggest risk associated with the mill because, "with a long history here, I have witnessed the damage from the last boom firsthand. Many men died way too young, and I don't care to watch a new generation be slowly poisoned to death." A Nucla resident wrote, "We are missing a whole generation of men in the area. This stems from the mining days of the past. I can't walk anywhere around Nucla without seeing someone pulling an oxygen bottle." Another noted, "Uranium isn't safe. . . . It's sacred ground to me out there, and we could find other ways to create employment without putting our people and land in danger." A Paradox respondent explained, "I'm against [the mill] because I want a safe and healthy place to live, without the possibility of the environment being contaminated for my family and I. For the short term, it could leave the community toxic and undesirable to live. . . . Who knows about the long-term effects?"

For activists shaping sites of resistance, justice and sustainability often mean combating the volatility and instability of extralocal extractive industries. As the founder of the Paradox Valley Sustainability Association founder said, "We started because we would like to see sustainable and environmentally and economically healthy jobs and development through the area. . . . We're not 'no resource extraction' advocates, but we've seen so much devastation and destruction from greed-based resource extraction in the past that this is what we really wanted to focus our time and energy on." Other opponents of uranium renewal have noted the striking differences between supporters' and opponents' notions of justice. A Nucla resident who is actively resisting the mill observed, "I think people whose parents were involved with uranium before, there seems to be a real identity with that and regardless of health impacts, they support it. . . . It was the boom of their life, the good times, and ever since, they've been in a bust economy because that's all they've wanted to do, and they are just waiting for the next uranium boom to happen." Trevor, a San Miguel County commissioner, made similar observations: "They feel like they helped stop World War II with Little Man and Fat Boy, and [they] bring that into today's issues. It is just so strange to me, and it is a psychological phenomenon that I suppose could be similar in nature to [the] Stockholm syndrome."

About 15 percent of survey respondents, many of them from Paradox, hinted at various social dislocations caused by the mill and the resulting community divisions. One of them said, "It is unfortunate the need for jobs has caused so much support for a project that could be so damaging—physically, culturally, and socially. Consider the controversy. Consider the fact that some people in opposition are unwilling to speak out because they fear they might lose their job or perhaps a lease. . . . This gangland mentality of 'like it or leave it' displayed by mill proponents is perhaps of greater consequence than realized. . . . Polarization in communities has already gone far beyond healthy debate."

Threats of violence and death and a repressive public space have widened already deep class- and identity-based divides. But activists in sites of resistance continue to engage in activities that reflect class status. Not only have they initiated highly publicized lawsuits against Energy Fuels and the Colorado Department of Public Health and Environment, but in August 2010 the Sheep Mountain Alliance brought the neo-hippie band Phish to Telluride for an anti-radiation, anti-mill concert. Though Heather would not reveal how the alliance is funded, she made it clear that her ability to execute long-term public campaigns is strengthened by the group's ample financial backing.

Spatial Poverty, Divergent Dislocations

The spatial isolation of uranium communities induces what I call *spatial poverty*, in which residents are significantly deprived of social networks that would help them alleviate their social dislocation, fight perceived environmental injustices, or explore alternative economic development options. Thus, social activism in uranium communities mobilizes in a more isolated context than it would in urban centers along the East or West coasts, and activists may not have access to institutional or organizational support or to a diverse array of perspectives on social justice issues. Spatial poverty interacts with persistent economic poverty to enhance class-based social divisions within and between regional communities, and this dynamic affects how tensions inform notions of environmental justice and mobilize divergent strains of activism.

Often, impoverished populations are typically the ones that create sites of resistance, rejecting industrial facilities and their risks after decades of unequal exposure to them. In most uranium communities, however, mobilized activists say, "Yes to the mill!" Even in Monticello, where sites of resistance dominate activism, ambivalent support for industry renewal exists. This betrays the ways in which uranium dependence enacts structural violence in uranium communities. Spatial and structural inequalities affect social interactions and allocation of opportunities in rural areas, constraining upward mobility and aiding poverty's regional persistence.[29] Residents embedded in persistently impoverished, spatially isolated, and natural resource–dependent communities are constrained to conceptualize notions of environmental justice that reflect what they *can* reasonably attain, given their real and spatial poverty: that is, autonomy over how they use their land and develop its wealth. Thus, social activism in sites of acceptance never really has a chance to instigate transformative social change. The activism that mobilizes in these spaces looks less like a thread of Polanyi's double movement and more like what I call the triple movement, in which activists mobilize to protect and encourage the free functioning of various markets systems (here, uranium markets) because they view these markets as part of their social fabrics and community identities. Their notions of environmental justice translate to reducing their experiences of social inequality by using the wealth embedded in their surroundings.

Communities such as Telluride and Ophir certainly grapple with spatial isolation and natural resource dependence, but they have parlayed that dependence into lucrative tourism-based economies. Thus, they have the privilege and the power to say, "No to the mill!". Because their spatial isolation does not equal spatial poverty, the notions of environmental justice that develop in these sites of resistance can be more transformative. Their mobilization is more akin to Polanyi's double movement, where activists target the state or otherwise try to reembed markets in their social-environmental contexts to provide people and communities with more protection from the risks of industrial production. Yet social sustainability remains elusive for the people and communities involved in this debate. They battle with one another in ways that create barriers to collaboration for community development across the region. Instead of assuaging social dislocation, the uranium industry exacerbates it, especially in terms of class-based tensions and spatial poverty.

6

"Better Regs" in an Era of Deregulation

Neoliberalized Narratives of Regulatory Compliance

Who is going to police the site for 1,000 years? I think this whole process has pointed out that regulations may be antiquated and not [keep] pace with knowledge. . . . It is very telling that the [EPA] is in the process of reviewing their regulations and revising them after twenty-five years, and we seem to be in a hell of a rush to build this mill now.
—Dave, Telluride town councilman

From the beginning, we have listened carefully to the public and worked with Energy Fuels to minimize risks to public health and the environment. [We enforce] strict environmental regulations [that] far exceed those in place when the last such mill was constructed more than twenty-five years ago.
—Chris Urbina, executive director, Colorado Department of Public Health and Environment

Colorado's Highway 141 winds between the resort town of Telluride and the uranium communities of Nucla and Naturita, and the sixty miles that separates them makes a world of difference. Landscapes and social contexts change simultaneously—from alpine mountain vistas to red rocks, from wealthy to working-class residents, from Uggs and Patagonia ski jackets to work boots and Wal-Mart coats. Each time I walk into the weathered visitor's center on Naturita's dilapidated Main Street, I am greeted by two or three friendly women. Their smiling faces betray the roughness inherent to life on Colorado's rugged Western Slope. This place has become one of my touchstones; I always check in when I'm in town. Brochures and maps fill the space, but my attention is always drawn to the bright orange sign with stark black letters plastered high on the wall behind the women, the sign that exclaims, "Yes to the Mill!"

During my first visit to the center, one volunteer delivered a well-spun narrative that has since echoed through my interviews, a narrative now familiar to you as well: "We've been involved with uranium for years. We know it better than other people do, especially those folks in Telluride. And it's safer now. The regs are better. And Energy Fuels is good to us because George Glasier is a local." When I visited the street's few open shops, residents and business owners were quick to share their positive perceptions of renewed production and defended their belief that better regulations now govern the industry. I soon learned my lesson: the mill is popular and powerfully defended, uranium is embedded in Nucla's and Naturita's social fabrics, and opposition is seen as irrational. Some activists in these sites of acceptance refused interviews because they were skeptical of an academic's interest in their community's energy development efforts. In fact, the editor of the *San Miguel Basin Forum* not only refused my invitations for an interview but published a critical and inaccurate editorial attacking my study.

On a mesa in the middle of vast wilderness, a few tiny communities are playing key roles in our nation's dilemma over energy policy and its complex links to economic security, poverty, sustainable energy, and resource dependence in rural areas. But the refrain I heard most often during my fieldwork focused on environmental and health regulations in the uranium industry. Residents' widely divergent perceptions of industry regulations and their enforcement marked a core distinction between activists mobilizing sites of acceptance and those in sites of resistance–that is, between the goals of activism in Polanyian double movements (which look to restrict and reembed free markets) and those in triple movements (which privilege free markets).

Political-Economic Trends in U.S. Environmental Regulations

Environmental regulations in the United States have a relatively short history. Though philosophers such as John Muir and Henry David Thoreau brought ideas of preservation and conservation into public discourse in the 1800s, national attention did not turn to the need for formal environmental regulations until 1963, when Rachel Carson's groundbreaking *Silent Spring* told the story of environmental and human health degradation related to pesticide exposures.[1] Her work emerged at a fortuitous time: with the oil spill in Santa Barbara and the Cuyahoga River in flames, environmental degradation due to decades of rampant industrialization was becoming increasingly apparent. These overlapping circumstances ushered in the first epoch of environmental policy in the United States.[2] Lasting from roughly 1969 through 1980, the era was not dominated by neoliberal logic but featured straightforward examples of Polanyi's double movement in action.

Environmental activists and policymakers of this era mobilized collective protest and targeted the federal government, urging it to create broad, far-reaching legislation to address industrial pollutants in air, water, and land. Under public pressure, Congress passed expansive environmental regulations, beginning with the National Environmental Protection Act of 1970, which stipulated that federal agencies had to formally assess federal projects' environmental effects while creating an avenue for public participation in these policy decisions.[3] In 1970 President Nixon established the EPA, combining all federal environmental research, regulatory, and enforcement efforts into one agency, which was charged with monitoring air, water, and land quality to protect human health.[4] By passing the Clean Air Act of 1970, the Clean Water Act of 1972, the Endangered Species Act of 1973, the Safe Drinking Water Act of 1974, and the Toxic Substances Control Act of 1976, Congress was responding to mobilized double movement activism, through which the public had demanded increased federal oversight of industrial pollutants and their environmental effects.[5]

As the decade's energy crisis deepened, corporate leaders claimed that these new regulations were keeping them from developing innovative energy sources. But despite the energy industry's attack on new environmental regulations, the federal government put even more protections into place. In the late 1970s, an environmental justice movement began to mobilize, with activism focused on the built environment, industrial pollutants, and protecting human health.[6] The movement included the Love Canal homemakers, who publicized the damage that industrial pollutants had done to people in their homes, churches, schools, and neighborhoods. Such mobilization efforts pressured Congress into passing the Superfund Act in 1980.[7]

In the 1980s, the U.S. approach to environmental policy began to change.[8] As neoliberal ideology became increasingly hegemonic, environmental legislation began to emphasize the management of natural resources and the efficient execution of environmental laws.[9] Attention turned to the marketplace, and environmental protection was increasingly perceived and framed as competing with economic prosperity.[10] Under President Reagan's administration, ideologies of deregulation and market-based logic were applied to all areas of government. Although Congress initially resisted this shift, the administration's control over the federal budget allowed the government to impose major budget cuts and thus reduce the enforcement capabilities of institutions such as the EPA.[11] Regulatory enforcement was crippled for years. Even when budgets increased under the Clinton administration, regulatory agencies could only rebuild enforcement capacities they had lost.[12] In this underfunded context, the National Environmental Protection Act and other regulatory processes became bureaucratic barriers based on cost-benefit ratios rather than the transformative environmental regulations the public had fought for in the previous decade.[13]

Today, concerns about sustainable development—a nebulous concept—drive modes of environmental regulation, but neoliberal logic remains entrenched and hegemonic.[14] Issues of sustainability interact with market-based logic, restricted budgets for regulatory enforcement, increased corporate self-monitoring, and environmental de- and reregulation.[15] Volumes of environmental regulations exist; but due to shrinking government budgets from the federal government all the way down to municipalities, only a fraction can be adequately enforced. So how can we live sustainably? How can we reconcile neoliberal discourse, which privileges free markets, with necessary environmental protections? How can we rethink the way in which we "incorporate into the building blocks of our economic activity in society—including the calculation of the gross national product—measures of environmental health, quality of life, and the full effects of human settlement patterns on the land and consumption of natural resources"?[16] How can we rethink market-based systems of land commodification?[17] The uneven and contested nature of environmental regulations has led to sharp critiques about the role of the federal government and corporations in sustainable environmental protection.[18] Some activists question corporate influence on environmental policy—for instance, the notoriously business-friendly Energy Policy Act of 2005, which granted industries such as nuclear power room to grow.[19] As we have seen, however, other activists privilege free markets above longer-term concerns about environmental integrity, due in large part to the ways in which these people and their communities have been impoverished by the neoliberal era's loss of social safety nets and increasing rates of U.S. inequality.

Uranium Regulations in the United States

Uranium regulations in the United States changed as uranium acquisition shifted from a government-run, monopsonistic, weapons-production market to a commercial, privatized, energy-production market. Before World War II, the Belgian Congo and Canada had provided most of the world's uranium ore, and the United States had imported most of its materials.[20] But when faced with wartime security threats, the nation began cultivating a domestic uranium market. In 1946, Congress passed the Atomic Energy Act (amended in 1954), legislating civilian rather than military control of nuclear technology, and established the AEC to oversee all nuclear activity in the United States.[21] AEC efforts to stimulate domestic uranium prospecting, mining, and milling produced rapid results. By 1955, the United States was the number-one producer worldwide.[22] During the 1960s bust, however, the AEC phased out fixed returns for uranium; and the government's stretch-out program formally ended in 1970.

The regulations established in the Atomic Energy Act were minimal at best (see chapter 2).[23] After the AEC-dominated era ended, public health and environmental concerns as well as the requirements of an expanding privatized nuclear power industry inspired a new series of regulatory legislation. Passed in 1974, the Energy Reorganization Act led to the establishment of the U.S. Department of Energy in 1974, the Nuclear Regulatory Commission in 1975, and the passage of the Uranium Tailings Radiation Control Act in 1978. Now the Nuclear Regulatory Commission became the top regulatory authority overseeing all uranium byproducts and tailings.[24] The Department of Energy took over nonregulatory functions such as nuclear power development, public education and outreach, and promotional efforts. As uranium's regulatory context changed, so did the role of the federal government. Because corporations trying to enter the market began asking for regulatory exemptions, Congress allowed the Nuclear Regulatory Commission to approve Agreement States, meaning that state agencies could become primary uranium regulators for nuclear facilities in their states if they could prove they had regulations and institutional enforcement capabilities and that were at least as stringent as the federal government's.[25] Currently, thirty-seven states, including Colorado and Utah, have Agreement State status.[26]

Colorado's Regulatory Context

The Colorado Department of Public Health and Environment has formal authority over the regulation of radioactive materials in the state. As I detailed in chapter 4, this means that permitting and regulating facilities such as the Piñon Ridge Uranium Mill fall within its purview.[27] In addition, the department oversees an assortment of industry processes, including facility licensure, inspections to assure environmental and safety compliance, security of uranium byproducts and tailings, improvement of regulations, and management of remediation records. Although the department has primary regulatory authority, other agencies monitor different aspects of uranium production, from air quality to water contamination. Thus, the department does not assess the cumulative effects of a facility such as Piñon Ridge or set overall standards. Further, there are no mechanisms to capture the cumulative effects of the increased industrial activity that a new mill facility would likely require, such as increased mining and trucking activities. Most of the monitoring at Piñon Ridge will be done by Energy Fuels Resources itself.

My interviews revealed that regulators are ambivalent about whether the lack of cumulative impact assessment and an increased need for corporate self-monitoring are problematic signs of deregulation, despite the industry's increased regulations since the 1970s. Tim, an engineer and an upper-level administrator in the department's Radiation Management Unit, said that while "there is nothing wrong with looking at comprehensive impacts, . . . this is not the way our government agencies are established."

He argued that people in specific agencies, such as the Air Quality Control Commission, are better equipped to monitor air quality than his agency is. However, some department employees do have concerns. Edward, an upper-level department administrator overseeing the Piñon Ridge application process, observed, "That's the problem, that no agency has authority to oversee . . . the overall effects of the mill in terms of extra mining, extra traffic, all that stuff. . . . You end up relying on the environmental impact analysis from those other guys and have to hope their information is correct." The question remains contentious among regulators and is a key issue for activists who are resisting construction of the mill.

Regulatory Performances on Record

Mill opponents mobilizing sites of resistance express general concerns about the regulatory records of the Colorado Department of Public Health and Environment and Energy Fuels and point to other cases in which the department's oversight has been publicly criticized. For example, it still maintains an active license for Cotter Corporation's Cañon City Uranium Mill outside Denver, although production has been halted there due to groundwater contamination caused by the facility.[28] The organization Colorado Citizens against Toxic Waste has repeatedly criticized the department's ability to enforce uranium industry regulations, asserting that it failed to notify the public for years about groundwater contamination and other pollution related to the Cotter Mill. In an official statement at a November 2012 public hearing about Piñon Ridge, Sharyn Cunningham, an activist with Colorado Citizens, said, "We know what it's like to live with the contamination from uranium mills and to work with short-staffed regulators. Unfortunately, those hearings were déjà vu for us who have real-world experience with Colorado radiation regulators. We don't want what happened to us to happen on the Western Slope."[29]

Other concerns revolve around Energy Fuels' efforts to operate and expand mines in Paradox Valley in anticipation of the Piñon Ridge Mill's construction. In the mining realm, the company's adherence to regulatory protocol has been less than exemplary. Because of violations at nearby Whirlwind Mine, the corporation received fourteen citations from the Mine Safety and Health Administration (MSHA) between March 19, and March 24, 2010. Although the company was fined only one hundred dollars per violation, the infractions were serious: failure to notify MSHA when opening a mine, lack of a ventilation plan, and electrical and machinery safety problems.[30] MSHA regulators have also documented violations in other Energy Fuels mines, including the La Sal Complex, Arizona #1, and Pinenut; and it reports that the company has been delinquent in paying fines related to these violations.[31] Further, because of its recent buyout of Denison Mines, Energy Fuels has absorbed and is currently contesting more than 180,000 dollars' worth of unpaid Denison fines, due in large part to worker deaths at Pandora Mine. In the same merger, Energy Fuels acquired the White Mesa Mill, whose safety record has been attacked by numerous activists, including Ute tribal members, who contend that leaks in the liners of tailings ponds have contaminated groundwater.[32]

Self-Monitoring in an Agreement State

At the Piñon Ridge Mill and in the local uranium mines that supply it, Energy Fuels will monitor its own daily compliance with environmental, safety, and health regulations related to uranium production. My interviews with state officials and company employees indicate that both institutions support self-monitoring, despite the practice's questionable outcomes in other contexts.[33] Tim asserted, "[Our department] specif[ies] the kind of monitoring that is required. . . . We can modify what Energy Fuels does. . . . Then we will monitor them when they're monitoring, and we will collect split samples from them. . . . There is a quality-control process that provides a check on both the lab and the sampling." Pete, an upper-level environmental officer with Energy Fuels, explained that maintaining professional codes of conduct is integral to adequate self-monitoring: "When half of our company is professional engineers, we have certain obligations to protect human health and environment. That's drummed into us. That's what we do as engineers."

Edward, a department administrator, observed that self-monitoring is built on trust: "There is an issue of scientific integrity, and you have to have a basic level of trust that the people collecting the data are honest and collecting good data." He noted that the company would not want to break its trust with Colorado's health department, given the resulting local impact: "It gets pretty personal in these small towns. . . . You can't fool around and mess up the data collection, because somebody's going to catch you right away." James, an Energy Fuels employee who collects baseline environmental data for the company's licensure application, told me, "My main concern right now is, we need to establish really good data. . . . We've done an excellent job picking a site, we've done an excellent job on our data collection, and the [department] is going to look at the science. They will have no choice but to issue a license because we've done everything they've asked as we've met with them." Former CEO George Glasier confirmed his faith in Energy Fuels' self-monitoring: "The [department] can come out randomly and check all of your monitoring devices. . . . We don't control the instruments, as they are calibrated by independent labs. . . . The state looks at automatic readings. . . . You can't really fool with these things. Plus, the [state has] tough regulators, some of the best in the nation."

Budget Constraints in an Agreement State

Activists in sites of resistance question Colorado's effusive support of Energy Fuels' ability to self-monitor. But in a neoliberal context, where devolution and shrinking public expenditures mean that states and municipalities are responsible for more regulations with fewer financial resources, perhaps such responses are necessary without being sufficient.

While Glasier may believe that the Colorado Department of Public Health and Environment is "the best in the nation" at regulating, budget cuts have seriously limited the department's capacity to enforce regulatory compliance. A main driver of these cuts is Colorado's Taxpayers Bill of Rights (TABOR), passed by residents in 1992.[34] Under TABOR provisions, revenues available to higher education, public health, and other government offices are severely restricted because local governments cannot raise taxes without voter approval. More importantly, they cannot spend revenues generated under current tax rates if those revenues exceed the rates of population growth and exchange.[35] In other words, if Colorado revenues grow faster than population rates or inflation do, the state cannot spend the money on government services but must refund individual taxpayers directly.

According to the Center on Budget and Policy Priorities, TABOR has "contributed to a significant decline in the state's public services."[36] Since the bill's passage, the state has refunded nearly 2 billion dollars to taxpayers, money that could have been spent on K–12 education, public health, and other social programs.[37] The spending cuts have deeply affected institutions such as the Colorado Department of Public Health and Environment and the Air Quality Control Commission. Not only have they gouged institutional capacities for regulatory enforcement and other preventive programs, but they have also impeded agencies' abilities to execute expensive and complex processes such as radiation and air-quality monitoring. All of the departmental employees I interviewed noted that Colorado's reregulated context interferes with their ability to visit sites and track corporations' self-monitoring, a problem exacerbated by uranium communities' spatial isolation. For instance, the Piñon Ridge site is 340 miles away from Denver, which would make regular trips to ensure regulatory compliance unfeasible.

Edward noted his agency's severe enforcement limitations: "We have a deep history and experience. . . . But state agencies are not always the best funded. I don't have a crew; I don't have unlimited money. I don't have *any* money. But we do the best we can because we have responsibilities as a public health organization to protect the public health. . . . We do collect our own samples, but not as often as we would like to because of the cost. We are trying to keep costs down." His colleague Tim echoed these observations about the department: "The complication [in enforcing regulations] is that the state government is very concerned about growth and doesn't want to approve twenty extra people to oversee a mill even though the company pays for it. . . . The catch is, the work doesn't change. We still have just as much work to do." Will, a public relations officer at the department, told me, "Self-monitoring is done because the state budget cannot support paying a bunch of government employees to go out and monitor without raising taxes. . . . You'll see real outrage then."

Faith in Regulatory Compliance

The Colorado Department of Public Health and Environment may have a limited budget and Energy Fuels may have various legal challenges, but most residents of Nucla and Naturita have faith that Colorado's regulations and their enforcement will be strict and that Energy Fuels is trustworthy. Trust in regulatory enforcement has helped nurture sites of acceptance among a majority of Paradox Valley residents. Luke, a Nucla resident, expressed support for uranium industry renewal because he believes regulatory compliance will be scrutinized: "These guys on the state level don't want anybody messing up their job, so they are going to see that these guys [Energy Fuels] are inspected and everything is up to date. Today, even in the mining industry, we are bombarded with paper, you know, safety. . . . The mill is the same way." Matthew, a Naturita resident, told me, "I think regulations will be well enforced for the [Piñon Ridge] mill in particular. . . . There are going to be so many people watching because it is the first conventional mill in thirty years. . . . Even if they didn't have the extra oversight, I don't think it would be a problem. It's so strictly managed and adhered to." Don, the influential Nucla pharmacist, agreed: "With regulations getting better and stronger, we know more. . . . I have no concern about the mill. . . . I know MSHA and [the Occupational Safety and Health Administration] check that fairly regularly." Lynne, a visitor's center volunteer and Naturita resident, asserted that mill regulations are much better than they were during previous booms: "From the time my dad was thirteen to the time he retired, they had already gotten strict with regulations. . . . They make sure everything is followed to the letter because they can't afford the penalties."

A survey I conducted in four Paradox Valley communities established that a critical mass of local residents actively support uranium renewal *because of* their faith that regulations will be adequately enforced: 75 percent agreed or strongly agreed that regulations governing the Piñon Ridge Mill are adequate to protect public and environmental health. Respondents felt knowledgeable about uranium regulations, and 76 percent said they knew "a lot" or "at least something" about industry regulations. When asked to identify risks associated with the mill, 62 percent said they believed the mill would pose no risks to their community or the local environment.

Public meetings also illustrated the central role that faith in regulations has played in mobilizing sites of acceptance. A Nucla resident attending a January 2010 public meeting said, "When my dad was in the mine, it was not a good situation. . . . We heard stories about how they pulled their t-shirts around their noses to keep the dust out. There was no, well, very little regulation. Well, we have regulations now, maybe too many." A valley resident at the same meeting noted, "People that are opposed will lead you to believe that we do things like we did thirty or fifty years ago. That's not the case. We've learned a lot of things that don't work, and now we have environmental laws and engineering controls."

Residents trust Energy Fuels to self-monitor and comply with regulations and trust Colorado officials to enforce those regulations. Tanya, the mayor of Naturita, said, "I truly believe the [department] will monitor to the best of their ability to make people as safe as they can make them. . . . The technology and the ability we have, . . . if anything starts to happen, before it even gets slightly hazardous, it will be taken care of." She also expressed strong trust in George Glasier: "[He] gave an awesome presentation to the chamber with a luncheon. [Energy Fuels] had tours on their site I went to and explained so many things in detail. I believe they have been an open book and try to get the public to understand everything they are doing. George has always answered questions." Her comments were echoed by most of the mobilized supporters whom I surveyed and interviewed. Paul, a thirty-year resident of Naturita, said, "Even though George's involvement in the mill process now is going to be curtailed with his retirement, I think he's a firm believer that these health regulations will be enforced, that he's not going to be running away from it. . . . You know, we all live here." Speaking of Energy Fuels, Paula, a Nucla resident, told me, "I have no reason not to trust them. They've been good community members here as far as being generous and helpful for community events. . . . They've done a good job, and I've heard they've been generous in making donations to the school." According to Rita, a Naturita resident, her strong trust in the company's ability to self-monitor is linked to its public presentations: "The first meeting I went to, there was a presentation by Energy Fuels about efforts to make sure the mill is using the most up-to-date information possible to be safe. . . . I appreciate that. It seems to me—what else can be done? Whatever is necessary, Energy Fuels is doing what it needs to do. When anybody asks them to comply with regulations, they always do it, and I've been impressed by that. George always answers our questions personally as well."

I heard versions of Rita's comments from a wide swath of residents, not only small business owners and community leaders but also working-class people; and those opinions were reflected on their survey responses as well. People mobilizing sites of acceptance trust the privatized uranium industry as well as current devolved regulatory structures. When I asked whom they trusted as the most important source of information about regulations affecting the mill, 77 percent identified the Colorado Department of Public Health and Environment as an important or very important source of information about the industry, mill safety, and regulations. A striking 78 percent reported that they had strong or complete confidence in the department's capacity to manage and regulate the Piñon Ridge Mill facilities and site, and 74 percent identified Energy Fuels as an important or very important source of trusted information about the industry and regulations. Abundant open-ended survey comments reinforced my perception of community-wide trust in these institutions.

Comments at various public meetings were consistent with those I heard in interviews and read on surveys. At a public meeting in January 2010, a longtime area resident observed, "We have state and government regulations that help keep us safe. . . . We are willing to trust those who have our safety at heart and best interest at heart, and we know you both [Energy Fuels and the department] will do a good job." Another Nucla resident said, "This is well regulated and we trust you guys to do that and make sure everybody here is safe." At a February 2010 meeting, a local chapter representative from the Pipefitters' Union expressed the union's trust in both institutions: "The rules and regulations that are put out by the [Nuclear Regulatory Commission] and these guys are phenomenal. My membership in the United Association has full confidence in Energy Fuels that this will be done right. We have full confidence in you guys [at the department] that you will make sure workers and the environment are protected." Another resident said, "We need jobs. We need to move our country toward energy independence. It can be done safely. With your help and oversight, [the department] can ensure that we achieve all of the above. Please approve the mill license."

Distrust in Regulatory Compliance

Opponents of renewed uranium production distrust industry regulations and question the state's enforcement capacity and Energy Fuels' compliance. According to Maureen, former program director of the Sheep Mountain Alliance, Colorado's oversight at other radioactively contaminated sites has been lax: "The remediation in Placerville, just down valley here, it's being overseen by [the Colorado Department of Public Health and Environment].[38] No people wear masks, there's no protection at all at that site. . . . [What I mean is] it's not that they don't have protection; it's that there's no enforcement. . . . I feel like the state laughs at the dangers of radiation." Mark, a hydrology professor at the University of Colorado–Boulder and a consultant for the Sheep Mountain Alliance, echoed her skepticism. He brought up the subject of the Summitville Gold Mine, where cyanide leaked into surrounding water sources: "[Summitville] continued to get permits from the state. . . . The state was watching. They misjudged the snow pack one year and the mine pit was filling up. . . . There was going to be a catastrophic release, and [the company] just walked away. They didn't shut the doors; they just walked, and cleanup costs today are over 200 million dollars." Mark worries that the department will hold Energy Fuels to the same low regulatory standards, especially given the agency's reduced budget and its reliance on the company to monitor its own regulatory compliance.

Public health specialist Doug Brugge has reservations about the quality and enforcement of regulations, especially given the Piñon Ridge Mill's remote location: "It's clear that existing regulations in the United States are inadequate. I'm also not sure that having the standards means that people will be protected. Standards get violated all the time, and enforcement can be weak. . . . So enforcement is the big issue, especially when you have a uranium mill out in a rural area." Heather, the president of the Sheep Mountain Alliance, is also concerned: "These people who say, 'Regulations are much better these days'—well, they aren't. . . . So saying, 'The state is going to protect us; the EPA is going to protect us'—well, no, they are not actually, and we need to continue to fight for those protections." Likewise, David, a Telluride town councilman, worries that long-term sustainability concerns cannot be addressed with current regulations because uranium mill tailings and mine wastes remain radioactive for thousands of years: "Who is going to police the site for 1,000 years? I think this whole process has pointed out that regulations may be antiquated and not [keep] pace with knowledge. . . . It is very telling that the [EPA] is in the process of reviewing their regulations and revising them after twenty-five years, and we seem to be in a hell of a rush to build this mill now."

Similar concerns emerged during regional public meetings, where audience members from outside Paradox Valley often mentioned the industry's expensive legacies and regulatory failures and the small amount of money Colorado requires Energy Fuels to set aside to pay for potential accidents and to decommission the mill. At a June 2010 meeting, a San Miguel County resident asserted, "Colorado taxpayers like myself have already spent over 1 billion dollars on cleaning up past uranium operations while much more cleanup lies ahead. Financial statements indicate that Energy Fuels cannot retain professional consultants to maintain the highest level of public health or environmental protection." A business owner in Montrose worried that "state regulations have not been and are not strong enough nor enforced well enough for the state to protect us from water contamination, air pollution, and health hazards." At another meeting, hosted jointly in June 2010 by the community of Ophir and the Colorado Department of Public Health and Environment, an Ophir resident alluded to the department's contested performance in regulating the Cotter Mill, stating that "the Cotter Corporation has a less-than-stellar record concerning their mill, and I expect the same here."

Though most of the people who responded to my survey expressed faith in regulatory compliance, 28 percent of respondents noted concerns about general safety risks and potential accidents. When I asked them about the adequacy of current uranium regulations, 15 percent disagreed or strongly disagreed with the idea that regulations were adequate. A Nucla resident explained that he opposed the mill because "regulations are circumvented every day, especially where corporations are concerned." A Paradox resident observed, "There is a false perception that regulations can make this mill safe. Regulations often set limits above which something is considered intolerably unsafe. . . . Common sense tells us that the only safe amount of emissions of this sort is zero."[39]

Mill opponents distrust Energy Fuels' claims of transparency, asserting that the regulatory department and the corporation have been too closely aligned during the licensure process. Clint, founder of the Paradox Valley Sustainability Association, wants a more democratic dialogue between various institutions and stakeholders at public meetings: "I don't believe we have had a detailed conversation about what regulations are, how they'll be enforced, and exactly what is going to make this situation different from the uranium boom that ended in 1979 or 1980. . . . Every time [opponents] get into a detailed conversation with the [Colorado Department of Public Health and Environment], we get the same response: 'We have your comment.' . . . There has never been a discussion or questions adequately answered." Though the November 2012 public hearing did create space for two-way communication, previous public meetings had already stoked distrust. For example, at a June 2010 meeting, a rabbi from Telluride declared, "The [department] is acting in a cavalier manner in the approach it's taking to protecting this region. . . . I have heard promises that monitoring will take place to make sure that the air and water are not contaminated, but these are hollow assertions . . . and mean nothing."

A vocal minority of survey respondents also expressed distrust. When asked why she saw the Sheep Mountain Alliance as an important source of information regarding the mill, a Nucla resident wrote, "I feel the different government and private entities need to be overseen. I don't have total faith in some of the regulatory agencies. I want to see 'watchdog' organizations included in sharing data." A Bedrock resident responded, "I do not believe that Energy Fuels can be trusted. And I hope [the department] can," while a Paradox resident wrote, "Too many questions have not been answered about emissions, groundwater, and land protection, long-term agricultural effects, and hiring from available personnel (poorly qualified by most accounts)."

Distrust in Energy Fuels' ability to monitor its own regulatory compliance has created fertile conditions for sites of resistance to mobilize. Mark the hydrologist told me that he rejects neoliberal logic about efficiency of self-monitoring: "Self-monitoring is worthless. . . . These companies are going to do things to serve their best interest. That is the nature of capitalism. That initial air-quality report that [Energy Fuels] put out, I looked at it and it was inadequate. . . . So they have already been busted once." Doug Brugge explained, "I think self-monitoring is a really bad idea. There's an economic disincentive because when you increase safety or environmental standards it costs money, which is why companies are resistant to do that. . . . If there's economic pressure to make more money and there's self-regulation of environmental controls, how's that going to play out?"

Jeanne, the wildlife biologist from Paradox, discussed her own laboratory experiences with self-monitoring: "It's like the fox guarding the hen house. There is just so much human nature that has to be depended upon in the whole monitoring process. With a high-risk thing like this, you can't afford them to do their own self-monitoring. . . . People do stupid things because they don't want to get in trouble or be bothered." Using the same metaphor, David the town councilman compared the potential chaos to Wall Street's disastrous experiences with self-monitoring: "It's like I have foxes that tend my hen houses. It works out great—I don't have any hens but don't have to clean up chicken poop. Investment banks have high regulations, are supposed to self-monitor. Did that work out very well in 2008?"

Lack of attention to cumulative effects also undergirds opponents' distrust. Mark explained, "One of my recommendations is that they look at cumulative impacts that include enhanced mining activities. The state has said no to that, which I find highly disappointing." Jeanne said, "There is no cumulative impact assessment, which is an integral part of the [National Environmental Protection Act] process. Their claim of what is a cumulative impact assessment is a joke." Caitlin, a member of Colorado's Air Quality Control Commission, lamented the lack of institutional mechanisms to facilitate cumulative impact assessments: "Our regulatory systems by statute are not set up to do the kind of broad overview. We have created silos so that we have only agencies that deal with air, water, radiation. . . . They don't talk to each other really well. . . . Since I've been on the Air Quality Control Commission, I've been yelling about it."

Opponents who attended public meetings noted the need for greater cumulative oversight, and many worried that corporate concerns about profit margins could interfere with regulatory compliance. One Telluride resident at a June 2010 public meeting said, "I don't want to take a chance on [Energy Fuels] breaking their promises. My sense is that they don't really care about clean air or water in the Piñon Ridge Mill. Their bottom line is the almighty dollar. They couldn't care less about you and me." Another attendee warned, "Don't overestimate the ability of [the Colorado Department of Public Health and Environment] or of Energy Fuels, or any other agencies and oversight [institutions] to monitor and enforce all the regulations. [Speaking to department officials:] Don't . . . make this venture another experiment with the American public and wildlife."

Regulatory Compliance and Structural Violence under Neoliberalism

Debates about uranium regulations and enforcement unfold in contexts shaped by neoliberal logic and governance, especially evident as de- and reregulation have repeatedly shifted the regulatory landscape governing uranium development. Even as national governments relinquish responsibilities to smaller political units such as states or municipalities, shrinking state budgets impede the enforcement capabilities of Agreement State institutions such as the Colorado Department of Public Health and Environment, which have become increasingly reliant on corporations to self-monitor for regulatory compliance.

Nonetheless, supporters of renewed uranium production mobilizing sites of acceptance maintain faith in regulations, even in devolved and reregulated contexts, and trust Energy Fuels and the department to monitor compliance. The structural positions of most industry supporters—particularly the persistent poverty and spatial isolation that have fueled dependence on uranium and other natural resource markets—constrain their communities' economic and other development choices. This sense of constraint has conditioned their trust in regulators and Energy Fuels and shifted their notions of environmental justice. Sites of acceptance such as these coalesce into larger triple movements, where activists privilege and protect free markets because they see them as part of their community's social fabrics and their own personal identities.

Meanwhile, opponents of renewed uranium production mobilizing sites of resistance are skeptical about regulatory compliance and distrust the department's and the corporation's ability to adequately protect public and environmental health. Most opponents are better positioned to challenge industry renewal than supporters are: their livelihoods and their communities' social fabrics are less directly dependent on renewal and are less constrained by historical connections to the industry. As such, activists mobilizing sites of resistance use more traditional notions of environmental justice and embody the double movement as they try to reembed uranium markets in social and environmental contexts, including the industry's underaddressed legacies.

The Piñon Ridge Uranium Mill has tested Colorado's devolved regulatory and enforcement capacities on two fronts. First, it has challenged the Colorado Department of Public Health and Environment's ability to supervise permitting, construction, and regulation of the first such facility built in three decades. Second, the permitting process has coincided with unprecedented austerity measures in the state and a chronic recession nationally. Even before the recession, neoliberal logic led to the passage of budget-slashing policies in Colorado, such as TABOR. As a result of such cuts, the department is now operating on a shrinking budget, though uranium renewal and mill construction specifically require the institution to have *increased* capacity to enforce regulations.

When a state institution becomes increasingly reliant on private corporations such as Energy Fuels to self-monitor regulatory compliance, it demonstrates its inability to fulfill its regulatory obligations as an Agreement State. So why do so many Nucla and Naturita residents trust this compromised and devolved system to adequately regulate uranium development? The answer resides in both class location and neoliberal logic. Social scientists David Harvey and Jeff Popke suggest that neoliberalism has such power because it connects to American notions of individual freedom and "instills [in people] an increasingly narrow and individualized sense of responsibility and ethical agency."[40] As the state retreats in de- and reregulated contexts, structural constraints require people to become more free, atomized, or self-governing, particularly in rural pockets that were originally settled as self-reliant economic outposts for resource extraction. Individualized work ethics and narratives of market-based self-sufficiency lead to a "remoralization of the poor," through which people feel solely responsible for economic and other successes or failures.[41] For economically vulnerable residents such as the working-class people and small business owners of Nucla and Naturita, a highly individualized sense of responsibility encourages market-based logic and trust in the capacity of devolved institutions to enforce public health regulations. When residents balance vivid economic need against the abstract long-term risks of uranium renewal, they find it more logical, rational, and realistic to privilege economic development over environmental or social considerations.

7

Conclusions and Solutions

Social Sustainability and Localized Energy Justice

A livelihood is socially sustainable which can cope with and recover from shocks, adapt to and exploit changes in its physical, social, and economic environment, and maintain and enhance capabilities for future generations.
—Robert Chambers and Gordon Conway, *Sustainable Rural Livelihoods: Practical Concepts for the 21st Century*

For years, I have watched the Victims of Mill Tailings Exposure fight a tireless battle for environmental and health justice. In our regular emails and phone calls and during my visits to Monticello, I have rejoiced in their small victories. But mostly I've felt devastated about their circumstances and powerless to help. I can only imagine how consistently helpless they feel, how constrained their options are. Social dislocation is palpable throughout the community; cancer clusters continue to emerge even as funding evaporates for temporary cancer screening and treatment facilities. Twenty years after first mobilizing, VMTE activists now consider taking the federal government to court, an option they have avoided due to limited time and financial resources. Yet many of these activists have also ambivalently accepted the revitalization of local uranium production—the very industry that shaped their economic instability and persistent poverty and damaged their daily quality of life.

Uranium communities such as Monticello offer important insights into the nuclear renaissance and its social sustainability, illustrating how production can perpetuate boom-bust economic cycles and lead to socio-economic volatility, inequality, and environmental degradation. As the VMTE's sustained fight for recognition highlights, uranium's commodification and extraction have created deep social dislocations and threatened social sustainability in these rural communities for decades. Yet as Nucla's and Naturita's histories demonstrate, activists' goals and people's notions of environmental justice depend on the historical, social, economic, and environmental contexts in which they are embedded. VMTE members' surprising acceptance of industry renewal reminds us that neoliberal hegemony can enact an especially pernicious form of structural violence in communities whose economies have adapted to producing natural resource commodities that are traded in privatized global markets and whose residents have little power to counter the norms of free-market privilege.

The complex historical legacies of previous uranium booms foreshadow how spatially isolated, persistently impoverished, and natural resource–dependent communities may fare in a renewed production boom. For Nucla and Naturita residents, construction of the Piñon Ridge Mill would make the possibility of economic security and industry revitalization more tangible, despite accompanying economic and environmental risks. Each of those towns has less than half the population of Monticello, along with far higher persistent poverty rates and spatial isolation. Mills and mines in and near those communities operated longer than the Monticello Mill did, so residents have a deeper identification with the industry—a bond fortified by the loss of nearby Uravan.

Nonetheless, Monticello's experiences are a cautionary tale. For nuclear power to be genuinely sustainable, communities that supply uranium and enable the fuel cycle should not be used as national sacrifice zones. Nuclear power should not be framed or funded as renewable energy until comprehensive government programs formally address its legacies in uranium communities and among segments of those communities not now covered under federal programs such as RECA, while addressing broader questions about its social sustainability. The sources of social dislocation I have pinpointed in uranium communities also plague other phases of nuclear energy production—for instance, power-plant construction and waste storage. As long as the industry's sustainability remains suspect, so does its ability to mitigate global climate change. Numerous leaders, policymakers, and citizens believe that social sustainability is an important component of renewable energy development. As former EPA director Lisa Jackson stated in her oath of office, "renewable energy should be able to cut through a thicket of thorny social ills and solve long-standing problems across the entire spectrum of American life."[1] Thus, the social and environmental costs in Monticello, Nucla, Naturita, and other uranium communities must be central to the discussion. The lessons we learn from these communities are key to finding solutions for socially sustainable energy development.

Material Conditions and Activism

Four material conditions are common to the uranium communities I have studied in this book: spatial isolation, persistent poverty, natural resource dependence, and environmental degradation, contamination and related health outcomes. The ways in which different activist groups experience and use these material conditions, especially as they relate to class position, mediate whether the groups mobilize as sites of acceptance or sites of resistance.

A key variable is personal and community identification with the uranium industry, which, for activists in sites of acceptance around Nucla and Naturita, interacts strongly with persistent poverty and is underscored by spatial isolation and natural resource dependence. Since the bust of the 1980s, local residents have watched their main streets decay and their populations dwindle. Yet they maintain a strong personal identification with the industry, believe they understand uranium better than most Americans do, and are not necessarily scared of it. Not only have they watched family members work in the industry, but it has given them a cultural connection to the natural landscape. These attitudes and conditions have influenced alternative notions of environmental justice, ones that center on local autonomy over land use as a way to diminish social inequality and persistent poverty by capitalizing on wealth embedded in surrounding landscapes. Yet this kind of activism is not particularly transformative in addressing social or economic inequality, protecting the environment, or shifting neoliberal power structures. Rather, as I've discussed throughout the book, these sites of acceptance contribute to a triple movement that privileges free markets above other socioeconomic concerns and entrenches the hegemonic power of neoliberal policy discourses in the United States.

For activists in sites of resistance, environmental contamination and concerns about natural resource dependence are the most salient material conditions structuring their experiences with the uranium industry. Sites of resistance tend to mobilize in places where inequitable environmental and health outcomes have more salience than persistent poverty or spatial isolation do—for instance, where environmental degradation has damaged health and well-being of marginalized populations or where, as in Telluride, economic stability requires an image of pristine nature. In Monticello, where a strong site of resistance to the industry's legacies does not erase ambivalent support for industry renewal, we see clearly how the industry's legacies are complicated by persistent poverty and spatial isolation. As such, activists' goals and their community histories interact in important ways with U.S. neoliberal hegemony, weakening the transformative potential of even the VMTE's activism.

In today's neoliberalized United States, commercial commodification of public land, the subsequent enclosure of uranium deposits, and privatized access to profit threaten to compromise rural residents' livelihoods. Gone are the days of prospecting and small-scale uranium trading, so common in the 1940s and 1950s. As the people of the Colorado Plateau have become embedded in global markets, they have become less integral to contemporary uranium extraction, which increasingly relies on labor-reducing mechanization. Yet Energy Fuels Resources, an international corporation that by strategic acquisition has become the largest conventional uranium producer in the United States, has cultivated its image as a local company. Through multiple modes of insinuation, it has influenced community land-use decisions and activists' goals, stirred up class-based tensions, and created division among stakeholders.[2] Even if Baobab Assets, led by Energy Fuels' former CEO George Glasier, purchases the permits and assets related to the Piñon Ridge Mill, these sites of acceptance are likely to remain energetically mobilized because supporters' trust in Energy Fuels has been based on their trust in Glasier as a local community member and ally.

As I discussed in chapter 4, the material conditions in sites of acceptance interact in important ways with the goals and image of Energy Fuels. In contrast, the material conditions in sites of resistance have created skepticism about the company's reliability and trustworthiness. Yet even among these activists, Energy Fuels' status as a private company complicates the situation. VMTE members, for instance, feel that corporations in this era might operate with fewer risks than the federal government did in previous eras, giving them hope that better regulatory enforcement might minimize environmental and health effects. However, this perception is not empirically supported, as the corporation's spotty environmental record at the White Mesa Mill demonstrates.

Reregulating Uranium

Most of the activists whom I interviewed or surveyed in sites of acceptance have faith in the industry regulations developed by the Colorado Department of Public Health and Environment as well as the agency's ability to enforce them. It seems that devolving governance—the shift of regulatory control from federal to state government—has generated local trust. Rural westerners have a long history of animosity toward federal land-use and natural resource policies, which has been complicated by their geographical distance from federal regulators, who are often located in distant East Coast cities. They are far more favorable to the state's local management teams, who appear often with Energy Fuels officials at public meetings about uranium renewal.

Activists in sites of resistance, however, do not believe that a private entity such as Energy Fuels should be trusted to self-monitor its compliance to industry regulations that opponents believe are inadequate to protect environmental integrity and the well-being of present and future generations. Activists from Telluride and from Monticello worry that the Colorado Department of Public Health and Environment does not have enough money to adequately enforce regulations due to budget cuts affecting the agency. They also fear that Energy Fuels' ability to self-monitor will be compromised by its main task as a corporation: to generate profits for shareholders. Penalties for regulatory infringements often mean fines for violators, so a strong conflict of interest arises when a public health department asks a corporate entity to report its own violations.

Activist organizations such as the Sheep Mountain Alliance distrust the state regulatory system so deeply that they have formed an agreement with Energy Fuels and the state of Colorado that allows them to conduct their own air-emissions and other tests to ensure the company's regulatory compliance. Here, even activists in sites of resistance are not challenging devolved governance, a signal of neoliberalism's hegemonic influence on social movements' tactics and goals.

Notions of Environmental Justice in Hegemonic Neoliberalism

Some scholars contend that the nation's shift to neoliberalism has fundamentally changed the transformative potential of activists and social movement organizations.[3] As these groups increasingly target corporations as the harbingers of social change, federal government authority and its perceived policymaking capacity are weakening.[4] According to analysts from the radical Right to the City Alliance, "transformative organizing works to transform the system, transform the consciousness of the people being organized, and in the process transform the consciousness of the organizer."[5] In a neoliberal context, however, with its new "social movement ecology," organizations are susceptible to co-optation, and their goals are diluted.[6] Neoliberalized settings facilitate "the ability of the corporate target to bring the interests of a challenging group into alignment with its own goals."[7] At stake is the very nature of social activism in the United States: can it enact transformative and fundamental social change when neoliberalism's values have become hegemonic modes of insinuation?

When activists in sites of acceptance who are constrained by material conditions such as poverty fail to challenge neoliberalism's core tenets, the transformative potential of their social activism is nullified because it fails to address the root inequalities structuring those material conditions. Even in sites of resistance, notions of environmental justice and forms of activism are not always deeply transformative. Organizations' strategies and goals follow market-based logic more than their revolutionary rhetoric might betray. Understandably, economic concerns influence their attitudes, but activists worry about more than the boom-bust cycle of instability. They also focus on how the uranium industry may hurt their own tourism- and recreation-based economies. In a chilly financial climate, all components of civil society are vulnerable to hegemonic norms, and rural livelihoods are at particular risk. Likewise, when practical economic goals must be prioritized over transformative goals, notions of environmental justice are often constrained or diluted. For instance, when faced with relatively austere funding for public healthcare, which is neither subsidized by the federal government nor readily accessible in the most isolated rural communities such as Monticello (even after passage of the Affordable Care Act), the VMTE shifted its attention to improving the availability of high-quality healthcare and coverage for treatment rather than concentrating on systemic changes to the uranium industry or broad-based, affordable healthcare for victims of environmental injustice. Tragically, the notions of justice and the modes of activism used by uranium community residents are completely rational in a neoliberal world.

The Triple Movement

As I have discussed, in the triple movement, activists use alternative notions of environmental justice that depend on market-based logic. Activists in sites of acceptance accept the risks of industrial production and trust the privatization and marketization of natural resources, seeing related markets as part of their social fabrics. They do not experience these markets as disembedded, so sites of acceptance contributing to the triple movement mobilize with fundamentally different goals from those of the sites of resistance that comprise the double movement. Their favorable perceptions of Colorado-based and corporate-monitored regulations and their distrust of the federal government contribute to triple movement mobilization as activists no longer feel the need to target the federal government to fight for the social protections that Polanyi predicted all activists would demand over time.

Today, with economic inequality at its highest level since 1928, such triple movements thrive throughout the United States among various types of social movement organizations and social concerns.[8] For instance, my ongoing fieldwork in communities associated with unconventional oil and gas production—especially where hydraulic fracturing methods are used during drilling—has uncovered strikingly similar patterns of mobilization. As in uranium communities, Pennsylvania residents have mobilized sites of acceptance and resistance in response to the rapid expansion of hydraulic fracturing in many of their rural communities, whereas activists and landowners in New York State have mobilized strong sites of resistance, passing a five-year state moratorium on hydraulic fracturing for natural gas.[9] In Weld County, Colorado, communities have equally divided into sites of acceptance and resistance. The county has become the most-drilled county in the United States, with strong sites of acceptance arising among its natural resource–dependent, impoverished populations and among populations who own mineral rights and benefit from that market. Many supporters even characterize this rapidly spreading production as "local energy," akin to local food. At the same time, sites of resistance have mobilized in relatively wealthy communities such as Fort Collins and Boulder, which have passed moratoria on unconventional drilling and are fighting for comprehensive environmental health assessments.[10]

This divergence in activism is evident across natural resource and energy development contexts. Coastal communities struggle with how to respond to climate change and lingering oil-spill contamination; Appalachian coal communities debate about how to expand a dying, high-pollution industry; Navajo Nation leaders vote to ban uranium production while they are also constrained to buy the Navajo Mine to supply their enormous coal-fired Four Corners Power Plant.[11] Similar sites of acceptance and resistance are mobilizing around renewable energy development, with wind, solar, and even geothermal approaches being alternately attacked and supported. I suspect that as market-based logic and individualized rights–based discourses become ever more hegemonic in the United States, questions of environmental justice will increasingly center on individuals' rights to control land-use decisions, whether or not those decisions are ecologically progressive or socially sustainable. Will the transformative potential of environmental justice activism be compromised over time, especially in contexts of rapid energy development? Perhaps not, but only if U.S. energy policy attends to transformative considerations such as social sustainability.

Social Sustainability and Renewed Uranium Production: The Small Scale of Energy Justice

According to Robert Chambers and Gordon Conway, a socially sustainable rural livelihood "can cope with and recover from shocks, adapt to and exploit changes in its physical, social, and economic environment, and maintain and enhance capabilities for future generations."[12] By this definition, then, uranium production is not socially sustainable for rural communities; and by extension, neither is nuclear power production. Uranium communities have not been historically resilient, nor is their resilience likely to strengthen in today's neoliberal climate. Residents have and will continue to endure chronic recession, underdevelopment, overadaptation, and rampant social dislocation due to persistent poverty as well as underaddressed environmental and health legacies. In addition, the nuclear fuel cycle creates more challenges than solutions for future generations. Most of us are aware of ongoing concerns about nuclear waste, but we have seen how uranium extraction—the front end of the nuclear fuel cycle—is also a large contributor to long-term toxic contamination and permanently unstable economies.

What are other options for energy and economic development in uranium communities throughout the Colorado Plateau? For the sake of fairness, I must first give the benefit of the doubt to renewed uranium production. Industry leaders such as Energy Fuels have the opportunity to learn from the mistakes of the past, to drive and structure a nuclear renaissance that is less environmentally, economically, and socially problematic. The proposed Piñon Ridge Mill may in fact turn out to be "the most technologically advanced and environmentally friendly mill in the world." That would be a terrific outcome. Uranium communities deserve a safe and healthy environment and stable economies, especially if they can actually gain some local control over how community land is used and how development is regulated. While such optimism is idealistic, it's not unreal.

Regardless of environmental or social outcomes during a third boom, we must immediately address the environmental, health, and economic legacies of previous uranium booms in these communities. This is particularly urgent if renewed uranium development does prove to be less destructive: communities and earlier generations of workers suffered tremendously as the government and the corporations made those initial mistakes. Monticello should not have to scramble to provide healthcare access or services for its residents, and those residents should not suffer from contested illnesses as, for example, 54 billion dollars in new nuclear reactor loans are preapproved. Before nuclear renaissance can be promoted as sustainable, the industry must take care of its legacies and of the communities and citizens who still suffer from them.

More than a thousand abandoned uranium mines dot the Colorado Plateau, and this estimate is conservative. As new research suggests, exposure to even low levels of uranium from abandoned mine shafts may lead to significant health problems, including kidney infections, endocrine disruption, and mutations such as breast cancer.[13] But there are practical approaches to dealing with those legacies while addressing the region's social dislocation. For instance, reclaiming and remediating abandoned mines would improve public and environmental health in the Four Corners.[14] Such efforts would require a significant labor force, perhaps from Nucla and Naturita residents who already say that they are willing and eager to work in the uranium industry. Of course, constraints such as reduced funding for social programs might limit their scope, but a public-private partnership between, say, the state of Colorado and the Telluride Foundation might make a remediation program a valid (if slightly neoliberalized) option for both economic development and public health.

Land grant universities, regional institutions such as the Western Rural Development Center, and smaller groups such as the Telluride Foundation are offering field-tested programs to sustainably revitalize rural economies in locally controlled ways. For example, the Western Rural Development Center facilitates community development and training programs that focus on positive assessments of assets in rural communities such as Nucla, Naturita, and Monticello. Alongside local leaders and residents, facilitators help plan economic development initiatives, holding multiday workshops where stakeholders meet to identify options. While these programs can suffer from unstable funding and endorse market-based approaches to development, their focus on community-based assets and collaboration makes them useful and effective models for introducing more sustainable development to uranium communities. With the mill construction delayed and its ownership changing hands, Nucla and Naturita residents have moved in this direction. Through the towns' shared chamber of commerce, they formed the West End Economic Development Corporation in 2013, which aims to "create an inviting, pro-business environment . . . while capitalizing on our natural and historic resources."[15] Though this approach is still highly market-based, it honors regional identification with uranium in less extractive, more tourism-based and even recreational modes of economic development.

The Colorado Plateau has abundant resources for wind, solar, and geothermal energy: the Four Corners region has more than three hundred days of sun per year, and wind patterns, especially in Utah, may make the area suitable for extensive wind power development. While these energy approaches carry their own social problems, technological limitations, and political baggage, they do represent a chance to separate current development efforts from the legacies of uranium production. Further, the implementation of regional energy projects would create collaborative spaces among otherwise spatially isolated and, at times, class-divided communities. This could lead to other innovative economic development initiatives as community leaders facilitate less competitive models for regional growth. For such collaboration to work, however, communities must overcome existing and substantial class-based tensions.

Some of these encouraging strategies are unfolding already. With uranium prices dropping and the industry slowing for now, Paradox Valley residents have made an alternative move: they have built one of the largest community-owned solar installations in the United States. The San Miguel Power Association, in partnership with the Clean Energy Collective, completed the 1.1 megawatt solar farm in late 2012.[16] Located very close to the uranium mill's proposed site, the farm has 4,680 solar panels and currently serves more than two hundred individual members.

If more energy systems and technologies were reduced to these smaller scales of production and distribution—at the regional, community, or even the neighborhood level—then we might be able to move away from large-scale, corporate-centered systems of raw material extraction and energy production. In neoliberalized systems of production, profit is privatized, and communities pay the environmental and social costs.[17] While it's important not to romanticize the local as a site of transformative social change, approaches such as the solar farm in Paradox may help communities such as Nucla and Naturita find workable solutions in which they may realize some genuine land-use autonomy and procedural forms of environmental justice. Smaller-scale energy production systems could take a variety of forms. They might look like the community-based, closed-loop systems now used in some Vermont industries, where waste streams from one economic sector provide biomass to produce energy in others. They may look like neighborhood-centered systems in which households share solar panels, wind turbines, and battery-storage capabilities, as in intentional communities such as Dancing Rabbit in Rutledge, Missouri. These systems require vision, technological prowess and investment, and a reorientation of energy-related research and development to focus on new solutions. Even a fraction of the 54 billion dollars in guaranteed loans for new nuclear reactors would go a long way toward developing these sorts of innovations. As long as we have the political will, alternative energy systems that are

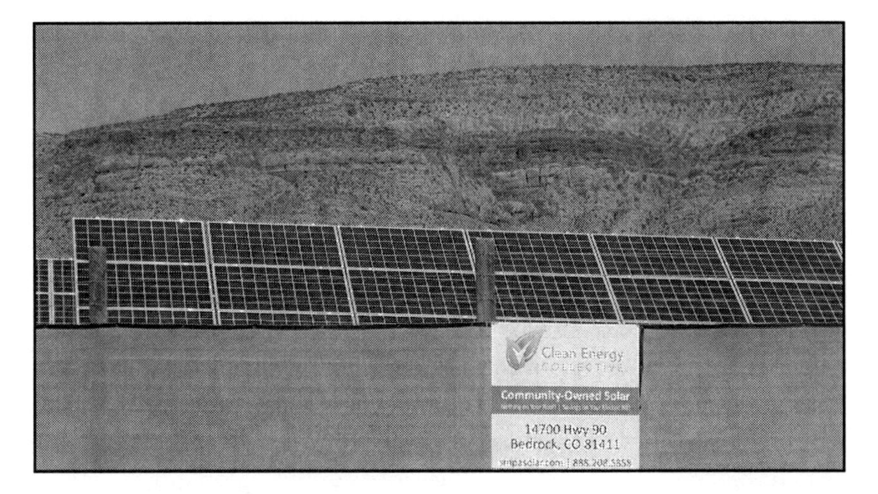

PHOTO 16 Community-owned solar operations offer a small-scale, community-based alternative to current energy development options being pursued in the Four Corners region. (Photo by Matthew Kazy)

socially and environmentally sustainable do not have to be limited to communes or science fiction.

By reorienting our energy production to economies of environments, or energy sheds, we could revolutionize how we produce and consume energy. If we linked energy production to neighborhoods, communities, and regions, people would gain a tangible understanding of energy's raw material inputs and its extractive phases, of how energy is produced, and of the meaning of massive American consumption patterns in those material contexts.[18] In scaled-down energy production systems, market considerations would drive development decisions in concert with social and environmental sustainability, even as markets would become a genuinely balanced part of community social systems. Scaled-down, community-based modes of energy production might also necessitate other scaled-down institutions that could provide cooperative rather than privatized social safety nets that are locally centered and locally provided. These community-based, community-scaled institutional arrangements would more sustainably address some of the key material and structural conditions plaguing uranium communities today: persistent poverty, spatial isolation, natural resource dependence, and environmental degradation.

To actualize these shifts, however, we need radical social activism: a mobilized citizenry that questions neoliberalism's hegemonic hold on U.S.

culture and agitates for transformative social change. But there must be space for this sort of activism. We need to empower activists to be truly transformative by creating institutional mechanisms and political spaces in which democratic discussions about energy, economic, and social policies can be productively and popularly debated. We need to dismantle neoliberalism's hegemony and re-create a society in which citizens do not internalize the belief that deregulated free markets will solve their daily struggles with poverty, inequality, and isolation. Bringing the scale of energy production and distribution down to the regional, community, or neighborhood level is just one means to this end. Only then will even our most remote, most impoverished communities realize true social justice, when they can achieve and enjoy sustainable rural livelihoods that will also be available to future generations. The stories of Fritz and Barbara Pipkin, their maps of community cancer clusters, the underaddressed legacies that lurk in thousands of living rooms across the Colorado Plateau: all of these realities should press us to create political spaces in which sites of resistance and larger double movements have the chance to flourish.

Appendix
Research Methods and Data Collection

I began conducting research on uranium communities when I was a master's candidate in sociology at Utah State University, continued my research during the rest of my graduate training, and am still focused on those places today. I feel lucky to be able tell their story because I believe that the histories and narratives of uranium communities and their residents are a vital yet hidden component of our national history and its energy policies. Following the lead of other scholars who practice public sociology, I have designed community-based examinations of environmental injustice, health concerns, social mobilization, corporate influence, and institutional responses related to uranium production and nuclear power. I work to ask questions that communities are interested in answering, in proposing problems that will be useful for them to solve. Using ethnographic and mixed methods, I conduct my research in the field as much as possible, with a focus on examining a multiplicity of household and community issues as they relate to larger production and political-economic systems.

To examine various types of social dislocation and social mobilization related to the Monticello cancer cluster, the Piñon Ridge Mill, and renewed uranium production, I use a number of mixed methods so that I can triangulate my data collection methods, strengthen the study's validity and reliability, and develop a more accurate perception of daily life in uranium communities.[1] For the purposes of this book, data include a variety of archival documents, in-depth interviews conducted between 2008 and 2011, ethnographic field notes, and a household survey instrument distributed in 2010 in the four communities closest to the mill site. I made several visits to uranium communities such as Monticello, Nucla, Naturita, and the abandoned Uravan sites, as well as to regional communities of interest, such as Telluride. Beginning in 2006, I collected similar data in Monticello, using the same ethnographic and mixed-method strategies.[2] My fieldwork in these Colorado and Utah uranium communities is still ongoing.

Archival documents have given me a context for understanding historical and current social, environmental, economic, and political issues surrounding uranium production and its renewal. They have also helped me identify key stakeholders and varieties of social activism. The research I present in this book relied on several forms of archival data: (1) public documents, such as those shared on corporate and government websites; (2) socioeconomic reports predicting the community and regional effects of uranium renewal, such as those compiled by county governments, corporate consultants, and nonprofit organizations; (3) licensure decisions and court documents released by state and federal agencies; (4) census and other demographic statistics; and (5) transcripts of public meetings. I examined each document at least three times, with the goal of selecting representative excerpts to identify and confirm thematic elements in the data. These later facilitated creation of interview questions, survey questions, and theoretical analysis.

To learn about daily experiences and quality of life in uranium communities, I conducted in-depth interviews with key stakeholders, whom I identified while I visited the region in August 2009 and during my initial analyses of archival data. I then used snowball sampling to identify other residents to interview in later phases of fieldwork.[3] Between September 2010 and March 2011, I conducted sixty formal, in-depth interviews. However, my fieldwork in uranium communities such as Monticello began in 2006. This durable relationship with the region and its communities allowed me to incorporate existing ethnographic research and interview data, which I enhanced with ongoing research in these communities. This has included numerous additional interviews since 2011 and a focus group in Monticello. I interviewed a wide variety of community residents, institutional representatives, activists, and corporate employees. Each interview was one to two hours long, usually in interviewees' homes or a place they had selected. A few long-distance interviews were conducted over the phone or via Skype, but all were recorded and then transcribed verbatim. I read each of those transcripts at least three times, working iteratively to identify and confirm themes, create coding schemes, and select representative quotations.

Between November 2010 and January 2011, I administered an Institutional Review Board–approved mail survey instrument to residents in the Colorado communities of Naturita, Nucla, Paradox, and Bedrock. Using a modified Dillman method, I created questions based on information gathered during archival analysis, community visits, and interviews, thus giving the survey local relevance and salience.[4] To compile baseline data and facilitate my ongoing fieldwork, I focused on eliciting opinions, asking respondents to identify community issues, levels of support or opposition to the Piñon Ridge Mill, perceptions of various state institutions and social movement groups, and benefits and risks associated with the mill. Questions also addressed the uranium industry's more general renewal. Open-ended questions let people elaborate on topics of particular personal concern. To facilitate a comparison of responses across residents and communities, I also tracked demographic variables such as age, income, education level, occupation, gender, and housing situation.

To create a sampling frame, I acquired address lists from the town clerks that included every homeowner or renter with a water, trash, or sewer bill. I confirmed names and addresses with the clerks and excluded repeated names as well as people with out-of-state addresses. (My focus was full-time rather than part-time residents.) Given the small number of households in these communities, I sent a survey to every household in Naturita, Paradox, and Bedrock.[5] Nucla was the only community with enough households (344) to make selecting a sample statistically worthwhile, so I randomly selected 278 households from my sampling frame.[6] After compiling the sample, I mailed the first wave of questionnaires to all recipients (N=520), asking the permanent adult householder with the most recent birthday to complete the survey. Two weeks later, I sent a follow-up postcard to households that had not yet responded. I distributed the second wave of surveys approximately two weeks after sending the follow-up postcard.

After removing undeliverables, I calculated the response rate at 41 percent: 212 out of 520 distributed surveys were returned. In Nucla, 168 of 278 households responded (60 percent); in Naturita, 77 of 204 responded (38 percent); in Paradox, 27 of 44 responded (60 percent); and in Bedrock 11 of 31 responded (35 percent).[7] To ensure that I had attained a representative sample despite the 41-percent response rate, I aggregated the data, which are comparable to 2000 census demographics for Nucla and Naturita (table A.1).

Slight differences do exist: survey respondents were older than the county population and included a higher percentage of males. After coding the survey data, I double-entered and analyzed them using the Statistical Program for the Social Sciences. I conducted primarily descriptive and comparative data analyses, such as t-tests and chi-square analyses, among the variables. In particular, I analyzed levels of support for the Piñon Ridge Mill, perceptions of regulations, and the level of trust in and importance of various institutions involved in permitting or regulating the mill.

My most recent research trips to uranium communities have included a guided tour of Uravan by members of the Rimrocker Historical Society, a self-guided walking tour of the Monticello Mill's remediated site and of memorials constructed by the VMTE, and a trip to the Piñon Ridge Mill's proposed site. I appreciate the warm welcome and openness I have typically experienced in these communities.

Table A.1. Comparisons between Survey and 2000 U.S. Census Demographics

Demographics	Survey Respondents	Nucla Census Data	Naturita Census Data
Age	59% ages 18–64 40% ages 65 and up	65% ages 18–64 13% ages 65 and up	62% ages 18–64 14% ages 65 and up
Sex	58% male 42% female	49% male 51% female	49% male 51% female
Occupation	15% agriculture/ ranching 9% production/ manufacturing 52% service sector	15% agriculture/ ranching 17% production/ manufacturing 67% service sector	21% agriculture/ ranching 16% production/ manufacturing 63% service sector
Income	17% less than $15,000 26% $15,000–34,999 19% $35,000–49,999 37% more than $50,000	26% less than $15,000 34% $15,000–34,999 17% $35,000–49,999 39% more than $50,000	24% less than $15,000 28% $15,000–34,999 23% $35,000–49,999 21% more than $50,000
Educational level	6% less than high school 36% high school diploma 28% some college, no degree 19% associate's or bachelor's degree	17% less than high school 40% high school diploma 29% some college, no degree 12% associate's or bachelor's degree	20% less than high school 54% high school diploma 20% some college, no degree 5% associate's or bachelor's degree

Notes

Chapter 1 Introduction: The Paradox of Uranium Production in a Neoliberal Era

1 Superfund sites, also known as National Priorities List sites, have been officially designated by the U.S. Environmental Protection Agency as the nation's most polluted and toxic sites, ones that warrant full remediation. Since its inception, the Superfund has been dealing with legal and budgetary challenges, so the program rarely adds new sites to its list. Monticello's radioactive and chemical contamination was so extensive, however, that two Superfund sites were designated in the town. See http://www.epa.gov/superfund.

2 Uranium tailings are the pulverized ore materials left over from the uranium milling process. Tailings are more radioactive than the ore is because their pulverization during processing allows radionuclides to escape into air and water more easily.

3 *Remediation* is the formal term for the process of cleaning and restoring highly contaminated land.

4 World Nuclear Association, "The Nuclear Renaissance," accessed June 5, 2014, http://www.world-nuclear.org/info/Current-and-Future-Generation/The-Nuclear-Renaissance.

5 For example, see http://healutah.org/nuclearutah/energy/greenriverreactors.

6 David Schlosberg began this thread of environmental justice theorizing in his inquiries into environmental justice movements. See his "Reconceiving Environmental Justice: Global Movements and Political Theories," *Environmental Politics* 13, no. 3 (2004): 517–540.

7 Stephanie A. Malin and Peggy Petrzelka, "Left in the Dust: Uranium's Legacy and Victims of Mill Tailings Exposure in Monticello, Utah," *Society and Natural Resources* 23, no. 12 (2010): 1187–1200.

8 Antonio Gramsci, *Selections from the Prison Notebooks*, ed. Quintin Hoare and Geoffrey Noel Smith (New York: International, 1971). Neoliberalism manifests itself via neoliberalization (the process of implementing neoliberal ideology via policy) in three major ways: a philosophy of free market superiority, a policy discourse

(including tenets such as privatization, marketization, state deregulation, market-friendly reregulation, use of market proxies in the state, civil society replacing the state, and creation of self-governing individuals), and a set of policy measures (such as free trade policies, education supporting marketization, social policies reflecting welfare-to-work paradigms, and governance policies such as devolution of federal responsibility to states and localities). See Noel Castree, "Neoliberalism and the Biophysical Environment 2: Theorising the Neoliberalisation of Nature," *Geography Compass* 4, no. 12 (2010): 1734–1746.

9 See chapter 4 for details about the mill's contested permitting process. It is also important to note that at the time of this book's publication, Energy Fuels had just entered in to talks with George Glasier's new Connecticut-based company, Baobab Asset Management, to purchase the permits and other assets connected the Piñon Ridge Mill, given the uranium market's recent weak behavior.

10 Karl Polyani created the concept of social dislocation, but this definition is mine because he did not provide one. See his *The Great Transformation: The Political and Economic Origins of Our Time* (Boston: Beacon, 1944).

11 See Telluride Foundation, "West End Community Resources," accessed June 5, 2014, http://www.telluridefoundation.org/index.php?page=west-end-building-community-resources.

12 Paul E. Farmer, *Pathologies of Power: Health, Human Rights, and the New War on the Poor*, 2nd ed. (Berkeley: University of California Press, 2004); Adam Burtle, "Structural Violence," accessed August 21, 2014, http://www.structuralviolence.org/structural-violence; Paul E. Farmer, Bruce Nizeye, Sara Stulac, and Salmaan Keshavjee. "Structural Violence and Clinical Medicine," *PLOS Medicine* 3, no. 10 (2006): e449.

13 Gregory Hooks and Chad L. Smith, "The Treadmill of Destruction: National Sacrifice Areas and Native Americans," *American Sociological Review* 69, no. 4 (2004): 558–575; Gregory Hooks and Chad L. Smith, "Treadmills of Production and Destruction: National Sacrifice Areas and Native Americans," *American Sociological Review* 69, no. 4 (2004): 558–575; Valerie Kuletz, "Appropriate/d Technology, Cultural Revival, and Environmental Activism," in *Appropriating Technology: Vernacular Science and Social Power*, ed. Ron Eglash, Jennifer L. Croissant, Giovanna diChiro, and Rayvon Fouche (Minneapolis: University of Minnesota Press. 2004), 287–305.

14 Allison Macfarlane, Michael Corradini, James K. Asselstine, and William G. Halsey, "Obstacles to Nuclear Power," *Bulletin of the Atomic Scientists* 63, no. 3 (2007): 24–25; World Nuclear Association, "The Nuclear Renaissance."

15 Bruce Finley, "If Colorado Yellowcake Uranium Mill Gets State's OK, Customers Would Likely Be in Asia," *Denver Post*, December 26, 2010, http://www.denverpost.com/ci_16943858.

16 U.S. Nuclear Regulatory Commission, "Combined License Applications for New Reactors," accessed June 5, 2014, http://www.nrc.gov/reactors/new-reactors/col.html; World Nuclear Association, "Nuclear Power in the USA," accessed June 5, 2014, http://www.world-nuclear.org/info/Country-Profiles/Countries-T-Z/USA—Nuclear-Power/#New_build. Importantly, these plants will take years to construct and operate, making the nuclear renaissance a slow-moving process.

17 For example, see Peter Hessler, "The Uranium Widows," *New Yorker*, September 13, 2010, 30–37.

18 Farmer introduced the term *structural violence* in his *Pathologies of Power.*

19 Susan Moran and Anne Raup, "Uranium Ignites 'Gold Rush' in the West," *New York Times*, March 28, 2007, http://www.nytimes.com/2007/03/28/business/28uranium.html.

20 "UXC Uranium Spot Price," *San Miguel Basin Forum*, March 3, 2011, 1.

21 The department is Colorado's top regulator of radioactive materials and the state's public health agency.

22 See, for example, Michael A. Amundson, *Yellowcake Towns: Uranium Mining Communities in the American West* (Boulder: University Press of Colorado, 2002). Importantly, this varies substantially from observations made by Bell and York about the ways in which corporate framing and public relations persuade rural residents to identify with the coal industry. While this is certainly a strong variable structuring community support and mobilized sites of acceptance related to uranium, I argue that the causal mechanisms are more complex and multifaceted. See Shannon Elizabeth Bell and Richard York, "Community Economic Identity: The Coal Industry and Ideology Construction in West Virgina," *Rural Sociology* 75, no. 1 (2010): 111–143.

23 To protect confidentiality, I have changed all names except when a person is a prominent public figure or an official with a public stake in the issues under discussion. In a lawsuit brought by Sheep Mountain Alliance and several co-plaintiffs, Energy Fuels was found to be in violation of the Atomic Energy Act for failing to allow a formal public hearing where residents could question regulators about the Piñon Ridge Mill and cross-examine witnesses under oath. Consequently, Energy Fuels held an official public hearing in November 2012, and the Colorado Department of Public Heath and Environment reviewed the mill's license and reissued permits. Recently, the U.S. Nuclear Regulatory Commission invited the public to review the procedures of Energy Fuels and the department.

24 Robert Chambers and Gordon Conway, *Sustainable Rural Livelihoods: Practical Concepts for the 21st Century*, IDS discussion paper 296 (Brighton, U.K.: Institute of Development Studies, 1991).

25 Lisa Jackson, cited in Douglas Fischer, "Clean Fuels Are a Social Panacea," *Daily Climate*, March 26, 2009, http://www.dailyclimate.org/tdc-newsroom/aspen/clean-fuels-to-the-rescue. Also see the U.N. Development Program reports on renewable energy, especially Amulya K. N. Reddy and U.N. Development Program, "Energy and Social Issues," in *Energy and the Challenge of Sustainability: World Energy Assessment* (New York: U.N. Development Program, 2000), 39–60.

26 U.S. Bureau of the Census, *Profile of General Population and Housing Characteristics: 2010*, http://www.census.gov/newsroom/releases/archives/2010_census/cb11-cn137.html.

27 Tom Power, *A Socioeconomic Analysis of the Impact of the Proposed Piñon Ridge Uranium Mill Project on Western Mesa, Montrose, and San Miguel Counties, Colorado* (Missoula, Mont.: Power Consulting, January 2011); U.S. Bureau of the Census, *Profile of General Population and Housing Characteristics: 2010.*

28 Power, *A Socioeconomic Analysis.*

29 Of the county's population, 83 percent is white, 13 percent Hispanic, and 6 percent Native American (U.S. Bureau of the Census, *Profile of General Population and Housing Characteristics: 2010).*

30 See note 25.

31 Ibid.

32 Ibid. Also see City-Data.com, "Telluride, Colorado (CO) Poverty Rate Data: Information about Poor and Low Income Residents," accessed August 21, 2014, http://www.city-data.com/poverty/poverty-Telluride-Colorado.html.

33 Energy Fuels, "Company Profile," accessed June 8, 2014, http://www.energyfuels.com/corporate/company_profile. Importantly, even if the mill's permit and related assets are sold to Baobab Asset Management, Energy Fuels' former CEO and honorary local George Glasier will once again be centrally involved in the mill's construction and related industry expansion.

34 Energy Fuels' Whirlwind Mine is located on the border of Colorado's Mesa County and the state of Utah. The U.S. Bureau of Land Management permitted the mine late last fall, though there have been some regulatory violations since it began operating. The Energy Queen Mine in Mesa County, Colorado, is another Energy Fuels property. In addition, the corporation has begun rehabilitation on the Tenderfoot Mine, located in San Juan County, Utah, just across the Colorado border. See Dick Kamp, "Uranium and Health: The Piñon Ridge Mill," *Montrose Daily Press Special Report*, November 16, 2008, http://uraniumwatch.org/pinonridgemill/uranium&health1.081116.pdf.

35 Energy Fuels, "Company Profile."

36 As I detail in chapter 6, Agreement States must meet certain criteria to show the Nuclear Regulatory Commission that they can regulate uranium activity at least as stringently as the commission itself would.

37 Nuclear Information and Research Service, letter from numerous organizations to the Colorado Department of Public Health and Environment, January 5, 2011, http://www.nirs.org/cdpheletter.pdf.

38 David Harvey, *A Brief History of Neoliberalism* (Oxford: Oxford University Press, 2005); Castree, "Neoliberalism and the Biophysical Environment."

39 Four dominant relations shape neoliberalism. *Governance* refers to the role of the state, which may devolve from the federal level to states or municipalities. Reregulation often accompanies state deregulation; here, the state becomes more of a market manager than a protector of citizens. Efforts include deregulating economic sectors such as energy and finance while reregulating others to ease market transactions. *Privatization* refers to the shift of natural resource ownership from the public sector to private firms or corporations and the shift from state-led regulatory and enforcement mechanisms to private ones. E*nclosure* aids privatization by co-opting common-pool resources such as land or uranium deposits under private ownership, typically excluding area communities from access. *Commodification* of complex, invaluable ecosystems creates markets for natural resources, enabling free trade of land, atmosphere, water, and goods such as uranium that were once noncommodified. See Noel Castree, "Neoliberalising Nature: The Logics of Deregulation and Reregulation," *Environment and Planning* A40, no. 1 (2008): 131; Castree, "Neoliberalism and the Biophysical Environment."

40 Castree, "Neoliberalising Nature: The Logics of Deregulation and Reregulation"; James McCarthy, "Commons As Counterhegemonic Projects," *Capitalism Nature Socialism* 16, no. 1 (2005): 9–24.

41 Nik Heynen and Paul Robbins, "The Neoliberalisation of Nature: Governance, Privatization, Enclosure and Valuation," *Capitalism Nature Socialism* 16, no. 1 (2005): 5–8.

42 For example, what neoliberal policies look like in uranium communities depends in part on historical contexts (see chapter 2).

43 Jamie Peck and Adam Tickell, "Neoliberalizing Space," *Antipode* 34, no. 3 (2002): 380–404.

44 Ibid., 389.

45 For example, corporate land appropriation and ownership lead to the privatization and enclosure of uranium resources that the state formally commodified less than a century ago. Valuation, or commodification, of uranium implicates communities in this enclosure effort in historically specific ways. As the industry privatized in the 1960s and 1970s, devolved governance and de- and reregulation of uranium markets reshaped the industry. Amendments to the Atomic Energy Act and the Energy Reorganization Act of 1974 played key roles in reregulating the U.S. uranium market to make it more amenable to global trade and private investment. The 1974 act split the U.S. Atomic Energy Commission into two entities: the Department of Energy would oversee development of nuclear weapons and power, and the Nuclear Regulatory Commission would regulate uranium and non-defense nuclear technology. The commission swiftly created the Agreement States program, further devolving nuclear regulatory oversight and governance to individual states that could prove their capacity to regulate uranium and nuclear industries within their borders. As of January 2012, thirty-seven states have entered into Agreement State status with the commission, taking over regulatory and enforcement responsibilities and governance but drawing from a shrinking pool of resources. Colorado became an Agreement State in 1966, which currently makes its Department of Public Health and Environment the primary regulatory authority over uranium. Yet like other states, Colorado has been compelled to slash budgets and public spending and has used neoliberal policy discourses to help normalize decisions to cut public services.

46 Jeff Popke, "The Spaces of Being In-Common: Ethics and Social Geography," in *The Sage Handbook of Social Geographies*, ed. Susan Smith, Rachel Pain, Sallie Marston, and John Paul Jones (London: Sage, 2010), 243.

47 Gramsci, *Selections*; Michel Foucault, *The Birth of Biopolitics: Lectures at the Collège de France, 1978–79*, ed. Michel Senellart, trans. Graham Burchell (Houndmills, U.K.: Palgrave Macmillan, 2008); Robert Fletcher, "Neoliberal Environmentality: Towards a Poststructuralist Political Ecology of the Conservation Debate," *Conservation and Society* 8, no. 3 (2010): 171–181; also see Julie Guthman, "Neoliberalism and the Making of Food Politics in California," *Geoforum* 39, no. 3 (2008): 1171–1183

48 In the 1970s, Foucault (in *The Birth of Biopolitics*) predicted the state of neoliberal governmentality.

49 Fletcher, "Neoliberal Environmentality," 175; Michael H. Finewood and Laura J. Stroup, "Fracking and the Neoliberalization of the Hydro-Social Cycle in Pennsylvania's Marcellus Shale," *Journal of Contemporary Water Research and Education* 147, no. 1 (2012): 72–79. Finewood and Stroup assert that neoliberal logic helps normalize hydraulic fracturing's effects on water quality in Pennsylvania.

50 Finewood and Stroup, "Fracking," 74.

51 Richard S. Krannich and Albert E. Luloff, "Problems of Resource Dependency in U.S. Rural Communities," in *Progress in Rural Policy and Planning*, ed. Andrew Gilg, David Briggs, Robert Dilley, Owen Furuseth, and Geoff McDonald (London: Bellhaven, 1991), 1:5–18.

52 See note 50; William R. Freudenburg, "Addictive Economies: Extractive Industries and Vulnerable Localities in a Changing World Economy," *Rural Sociology* 57, no. 3 (1992): 305–332; William R. Freudenburg and Robert Gramling, "Linked to What? Economic Linkages in an Extractive Economy," *Society and Natural Resources* 11, no. 6 (1998): 569–586; William R. Freudenburg and Scott Frickel, "Digging Deeper: Mining-Dependent Regions in Historical Perspective," *Rural Sociology* 59, no. 2 (1994): 266–288.

53 Freudenburg, "Addictive Economies," 306.

54 See the literature review and ongoing academic debate in Michael D. Smith, Richard S. Krannich, and Lori M. Hunter, "Growth, Decline, Stability, and Disruption: A Longitudinal Analysis of Social Well-Being in Four Western Rural Communities," *Rural Sociology* 66, no. 3 (2001): 425–450.

55 Craig R. Humphrey et al., "Theories in the Study of Natural Resource–Dependent Communities and Persistent Rural Poverty in the United States," *Persistent Poverty in Rural America*, special issue (1993): 136–172; David Louis Brown, Louis E. Swanson, and Alan W. Barton, eds. *Challenges for Rural America in the Twenty-First Century* (University Park: Penn State University Press, 2003); Debra Lyn Bassett, "Distancing Rural Poverty," *Georgetown Journal on Poverty Law and Policy* 13 (2006): 3–32.

56 Humphrey et al., "Theories"; Nancy Lee Peluso, Craig R. Humphrey, and Louise P. Fortmann, "The Rock, the Beach, and the Tidal Pool: People and Poverty in Natural Resource–Dependent Areas," *Society and Natural Resources* 7, no. 1 (1994): 23–38.

57 See Bassett, "Distancing Rural Poverty." On internal colonialism, see Pablo Gonzalez Casanova, "Internal Colonialism and National Development," *Studies in Comparative International Development* 1, no. 4 (1965): 27–37; and Peluso et al., "The Rock, the Beach, and the Tidal Pool."

58 David S. Walls, "Internal Colony or Internal Periphery? A Critique of Current Models and an Alternative Formulation," in *Colonialism in Modern America: The Appalachian Case*, ed. Helen M. Lewis, Linda Johnson, and Don Askins (Boone, N.C.: Appalachian Consortium Press, 1978): 319–349; Scott Frickel and William R. Freudenburg, "Mining the Past: Historical Context and the Changing Implications of Natural Resource Extraction," *Social Problems* 43, no. 4 (1996): 444–466; Bassett, "Distancing Rural Poverty."

59 Neil Smith, *Uneven Development: Nature, Capital, and the Production of Space* (Athens: University of Georgia Press, 2008), 206.

60 See Valerie Kuletz, *The Tainted Desert: Environmental Ruin in the American West* (New York: Routledge,1998), xiv; and Arn Keeling, "'Born in an Atomic Test Tube': Landscapes of Cyclonic Development at Uranium City, Saskatchewan," *Canadian Geographer* 54, no. 2 (2010): 228–252. Keeling shows how natural resource dependence and state-led uranium development led to similar outcomes in Canadian rural spaces.

61 For example, Nucla and Naturita have been less able to diversify, leading to unidimensional local economies and peripheral relationships to production centers. Population decline also plagues natural resource–dependent peripheries when their main extractive industry busts, as evidenced by Nucla and Naturita's combined 14 percent drop in population since 2000. See Power, *A Socioeconomic Analysis*; U.S. Census Bureau, "Nucla Town, CO" and "Naturita Town, CO," *American Fact*

Finder, accessed October 11, 2010, http://factfinder2.census.gov/faces/nav/jsf/pages/index.xhtml.

62 Richard C. Stedman, "Places of Escape: Second-Home Meanings in Northern Wisconsin, USA," in *Multiple Dwelling and Tourism: Negotiating Place, Home, and Identity*, ed. Norman McIntyre, Daniel Williams, and Kevin McHugh (Wallingford, U.K.: CABI Press, 2006), 129–144.

63 Francis T. Achana and Joseph T. O'Leary, "The Transboundary Relationship between National Parks and Adjacent Communities," in *National Parks and Rural Development: Practice and Policy in the United States*, ed. Gary E. Machlis and Donald R. Field (Washington, D.C.: Island Press, 2000), 67–87. Overall, research on tourism-related resource dependence suggests that community outcomes are typically negative or mixed. See Richard S. Krannich and Peggy Petrzelka, "Tourism and Natural Amenity Development: Real Opportunities," in Brown et al., *Challenges for Rural America*, 190–199.

64 Humphrey et al., "Theories"; Brown et al., *Challenges for Rural America*.

65 Paul Mohai, David Pellow, and J. Timmons Roberts, "Environmental Justice," *Annual Review of Environment and Resources* 34 (2009): 406.

66 Julie Sze and Jonathan K. London, "Environmental Justice at the Crossroads," *Sociology Compass* 2, no. 4 (2008): 1331–1354.

67 David Schlosberg, "Theorising Environmental Justice: The Expanding Sphere of a Discourse," *Environmental Politics* 22, no. 1 (2013): 44.

68 Ibid., 51.

69 Bullard, "Solid Waste Sites and the Black Houston Community," *Sociological Inquiry* 53, nos. 2–3 (1983): 273–288; United Church of Christ Commission for Racial Justice, *Toxic Wastes and Race in the United States: A National Report on the Racial and Socio-Economic Characteristics of Communities with Hazardous Waste Sites* (New York: Public Data Access, 1987); Andrew Szasz and Michael Meuser, "Environmental Inequalities: Literature Review and Proposals for New Directions in Research and Theory," *Current Sociology* 45, no. 3 (1997): 99–120.

70 U.S. General Accounting Office, *Siting of Hazardous Waste Landfills and Their Correlation with Racial and Economic Status of Surrounding Communities* (Gaithersburg, Md.: U.S. General Accounting Office, 1983); Bullard, "Solid Waste Sites"; United Church of Christ Commission, *Toxic Wastes*; Bunyan I. Bryant and Paul Mohai, *Race and the Incidence of Environmental Hazards* (Boulder, Colo.: Westview, 1992); Bullard, *Confronting Environmental Racism: Voices from the Grassroots* (Cambridge, Mass.: South End, 1993); Robert D. Bullard, "A New 'Chicken-or-Egg' Debate: Which Came First—The Neighborhood, or the Toxic Dump?" *Workbook* 19, no. 2 (1994): 60–62; Richard Hofrichter, *Toxic Struggles: The Theory and Practice of Environmental Justice* (Salt Lake City: University of Utah Press, 1993).

71 Laura Pulido, "A Critical Review of the Methodology of Environmental Racial Research," *Antipode* 28, no. 2 (1996): 142–159; Szasz and Meuser, "Environmental Inequalities"; Andrew Hurley, *Environmental Inequalities: Class, Race, and Industrial Pollution in Gary, Indiana, 1945–1980* (Chapel Hill: University of North Carolina Press, 1995). Hurley analyzed U.S. Steel's historical presence in Gary, the exposure of local minority populations to environmental toxins, and their activism against the company's industrial pollution.

72 Barbara L. Allen, *Uneasy Alchemy: Citizens and Experts in Louisiana's Chemical Corridor Disputes* (Cambridge, Mass.: MIT Press, 2003); Steve Lerner, Diamond: *A*

Struggle for Justice in Louisiana's Chemical Corridor (Cambridge, Mass.: MIT Press, 2005); Malin and Petrzelka, "Left in the Dust"; Hooks and Smith, "The Treadmill of Destruction"; Hooks and Smith, "Treadmills of Production"; Noriko Ishiyama, "Environmental Justice and American Indian Tribal Sovereignty: Case Study of a Land-Use Conflict in Skull Valley, Utah," *Antipode* 35, no. 1 (2003): 119–139; Valerie Kuletz, "Invisible Spaces, Violent Places: Cold War Nuclear and Militarized Landscapes," in *Violent Environments*, ed. Nancy Lee Peluso and Michael Watts (Ithaca, N.Y.: Cornell University Press, 2001), 237–260; Pulido, "A Critical Review"; Jill Harrison, *Pesticide Drift and the Pursuit of Environmental Justice* (Cambridge, Mass.: MIT Press, 2011).

73 Mohai et al., "Environmental Justice," 406.

74 Sze and London, "Environmental Justice at the Crossroads."

75 For examples, see Daniel Faber and Deborah McCarthy, "Neo-Liberalism, Globalization, and the Struggle for Ecological Democracy: Linking Sustainability and Environmental Justice," in *Just Sustainabilities: Development in an Unequal World*, ed. Julian Agyeman, Robert D. Bullard, and Bob Evans (Cambridge, Mass.: MIT Press, 2003), 38–63; Schlosberg, "Reconceiving Environmental Justice"; and Schlosberg, "Theorising Environmental Justice." Also see Daniel Jaffee, *Brewing Justice: Fair Trade Coffee, Sustainability, and Survival* (Berkeley: University of California Press, 2007), which looks at fair trade coffee markets through a neoliberal lens.

76 Guthman, "Neoliberalism"; J. Timmons Roberts and Melissa M. Toffolon-Weiss, *Chronicles from the Environmental Frontline* (Cambridge, U.K.: Cambridge University Press, 2001).

77 Roberts and Toffolon-Weiss, *Chronicles*; Mohai et al., "Environmental Justice"; Schlosberg, "Theorising Environmental Justice."

78 Dorceta E. Taylor, "The Rise of the Environmental Justice Paradigm: Injustice Framing and the Social Construction of Environmental Discourses," *American Behavioral Scientist* 43, no. 4 (2000): 508–580.

79 David N. Pellow, *Garbage Wars: The Struggle for Environmental Justice in Chicago* (Cambridge, Mass.: MIT Press, 2002).

80 Phil Brown, *Toxic Exposures: Contested Illnesses and the Environmental Health Movement* (New York: Columbia University Press, 2007), 1.

81 Contested Illnesses Research Group, *Contested Illnesses: Citizens, Science, and Health Social Movements*, ed. Phil Brown, Rachel Morello-Frosch, and Stephen Zavestoski (Berkeley: University of California Press, 2012); Brown, *Toxic Exposures*; Phil Brown and Edwin Mikkelsen, *No Safe Place: Toxic Waste, Leukemia, and Community Action* (Berkeley: University of California Press, 1990).

82 Movements against neoliberalization have dealt with water privatization and delivery, forest management, mining, and inequality in agriculture. See Karen Bakker, *Privatizing Water: Governance Failure and the World's Urban Water Crisis* (Ithaca, N.Y.: Cornell University Press, 2010); Karen Bakker, *An Uncooperative Commodity: Privatizing Water in England and Wales* (Oxford: Oxford University Press, 2003); Thomas Perreault, "From the Guerra del Agua to the Guerra del Gas: Resource Governance, Neoliberalism, and Popular Protest in Bolivia," *Antipode* 38, no. 1 (2006): 150–172; Scott Prudham, "Poisoning the Well: Neoliberalism and the Contamination of Municipal Water in Walkerton, Ontario," *Geoforum* 35, no. 3 (2004): 343–359; James McCarthy, "Devolution in the Woods: Community Forestry As Hybrid Neoliberalism," *Environment and Planning* A37, no. 6 (2005): 995–1014;

Nik Heynen and Harold A. Perkins, "Scalar Dialectics in Green: Urban Private Property and the Contradictions of the Neoliberalization of Nature," *Capitalism Nature Socialism* 16, no. 1 (2005): 99–113; Jeffrey Bury, "Livelihoods in Transition: Transnational Gold Mining Operations and Local Change in Cajamarca, Peru," *Geographical Journal* 170, no. 1 (2004): 78–91; Sandy Brown and Christy Getz, "Privatizing Farm Worker Justice: Regulating Labor through Voluntary Certification and Labeling," *Geoforum* 39, no. 3 (2008): 1184–1196; Jill Harrison, "Abandoned Bodies and Spaces of Sacrifice: Pesticide Drift Activism and the Contestation of Neoliberal Environmental Politics in California," *Geoforum* 39, no. 3 (2008): 1197–1214.

83 McCarthy, "Commons As Counterhegemonic Projects," 995.

84 Becky Mansfield, "Privatization: Property and the Remaking of Nature–Society Relations," *Antipode* 39, no. 3 (2007): 393–405.

85 Gabriela Valdivia, "On Indigeneity, Change, and Representation in the Northeastern Ecuadorian Amazon," *Environment and Planning* A37, no. 2 (2005): 285–303.

86 Warwick E. Murray, "The Neoliberal Inheritance: Agrarian Policy and Rural Differentiation in Democratic Chile," *Bulletin of Latin American Research* 21, no. 3 (2002): 425–441.

87 Ryan Holifield, "Neoliberalism and Environmental Justice in the United States Environmental Protection Agency: Translating Policy into Managerial Practice in Hazardous Waste Remediation," *Geoforum* 35, no. 3 (2004): 285–297; Guthman, "Neoliberalism," 1171.

88 For example, federal public health agencies such as the U.S. Agency for Toxic Substances and Disease Registry have found elevated rates of cancer but not a statistically significant elevation that they can officially connect to the Monticello Uranium Mill. Studies in other uranium communities find significant increases in lung cancer among miners but no evidence of excess deaths due to uranium exposure among female residents or mill workers. Recent studies record links between low-level uranium exposure and endocrine disruption and between uranium mining and air, water, and soil contamination. Toxicologists also have mounting evidence that uranium can have "adverse effects on the brain, on reproduction, on gene expression, and on mine workers and radon exposure" (Doug Brugge and Virginia Buchner, "Health Effects of Uranium: New Research Findings," *Reviews on Environmental Health* 26, no. 4 [2011]: 231–249).

Further, while federal legislation such as the Radiation Exposure Compensation Act, as amended in 2000, has compensated many uranium industry employees for their illnesses, community members who did not work in the industry have not been protected. Illnesses related to environmental uranium exposure remain contested by the state as well as by many activists who support uranium industry renewal. Few studies examine impacts to residential populations living near uranium production facilities. See U.S. Agency for Toxic Substances and Disease Registry, *Preliminary Public Health Assessment for Monticello Mill Tailings [DOE] and Monticello Radioactively Contaminated Properties [aka Monticello Vicinity Properties], Monticello, San Juan County, Utah* (Atlanta: U.S. Department of Health and Human Services, 1997).

89 See reviews in Beth Schaefer Caniglia and JoAnn Carmin, "Scholarship on Social Movement Organizations: Classic Views and Emerging Trends," *Mobilization* 10, no. 2 (2005): 201–212; Polanyi, *The Great Transformation.*

90 Daniel Jaffee, "Fair Trade Standards, Corporate Participation, and Social Movement Responses in the United States," *Journal of Business Ethics* 92, no. 2 (2010): 267–285; Daniel Jaffee, "Weak Coffee: Certification and Co-optation in the Fair Trade Movement," *Social Problems* 59, no. 1 (2012): 94–116; Tim Bartley, "Institutional Emergence in an Era of Globalization: The Rise of Transnational Private Regulation of Labor and Environmental Conditions," *American Journal of Sociology* 113, no. 2 (2007): 297–351.

91 Jaffee, "Weak Coffee," 112. See also Andy Szasz, *Shopping Our Way to Safety: How We Changed from Protecting the Environment to Protecting Ourselves* (Minneapolis: University of Minnesota Press. 2010).

92 Alexa J. Trumpy, "Subject to Negotiation: The Mechanisms behind Co-Optation and Corporate Reform," *Social Problems* 55, no. 4 (2008): 480–500.

93 Brayden G. King and Nicholas A. Pearce, "The Contentiousness of Markets: Politics, Social Movements, and Institutional Change in Markets," *Annual Review of Sociology* 36 (2010): 256.

94 Edward T. Walker, Andrew W. Martin, and John D. McCarthy, "Confronting the State, the Corporation, and the Academy: The Influence of Institutional Targets on Social Movement Repertoires," *American Journal of Sociology* 114, no. 1 (2008): 44.

95 Eric Mann, *Playbook for Progressives: 16 Qualities of the Successful Organizer* (Boston: Beacon, 2011), x.

96 Polanyi, *The Great Transformation*, 71.

97 Ibid., 79.

98 Progressive counter-movements agitate for structural change and state protections to quell social dislocation. Regressive movements are more reactive, retracting into familiar cultural frameworks in response to social dislocation. Conceptually and practically, both movements attempt to protect societies from the chaos of disembedded market systems. Fiercely neoliberal forms of free market movements emerged in the 1980s and became less visible and more hegemonic over time. Social scientists continue to examine the emergence of Polanyian counter-movements, the role of the state, and the need for political space for movement emergence. As the state retreats under neoliberal reregulation, democratic space becomes even more vital for counter-movement emergence. The concern becomes who provides that space. Empirical evidence is limited but does suggest that counter-movement mobilization relies on the existence of democratic social space because economic elites and affiliated political officials have fluid, abundant resources to purchase or commandeer political and ideological space if necessary. Recent inquiries into Polanyian counter-movements in neoliberal contexts tend to focus on progressive versions, even portraying the Mafia in 1990s Russia as a protective, progressive counter-movement. Such limited or confused interpretations lead scholars to interpret nationalist, imperialist, and even corporatist strategies in a neoliberal context as progressive counter-movements aimed at re-embedding economic markets in society. See Harvey, *A Brief History*; Fred Block, "Polanyi's Double Movement and the Reconstruction of Critical Theory," *Revue Interventions Économiques*, no. 38 (2008), http://interventionseconomiques.revues.org/274; Beverly J. Silver and Giovanni Arrighi, "Polanyi's 'Double Movement': The Belles Époques of British and U.S. Hegemony Compared," *Politics and Society* 31, no. 2 (2003): 325–355; Ayşe Buğra and Kaan Ağartan, *Reading Karl Polanyi for the Twenty-First Century: Market Economy As a Political Project* (New York: Palgrave Macmillan, 2007); Gareth Dale,

"Social Democracy, Embeddedness, and Decommodification: On the Conceptual Innovations and Intellectual Affiliations of Karl Polanyi," *New Political Economy* 15, no. 3 (2010): 369–393; Björn Hettne, "Security and Peace in Post–Cold War Europe," *Journal of Peace Research* 28, no. 3 (1991): 279–294; Robert W. Cox, "The Crisis in World Order and the Challenge to International Organization," *Cooperation and Conflict* 29, no. 2 (1994): 99–113; Sally Randles, "Issues for a Neo-Polanyian Research Agenda in Economic Sociology," in *Karl Polanyi: New Perspectives on the Place of the Economy in Society*, ed. Mark Harvey, Ronnie Ramlogan, and Sally Randles (Manchester, U.K.: Manchester University Press, 2007): 409–434; Fikret Adaman, Pat Devine, and Begum Ozkaynak, "Reinstituting the Economic Process: (Re)embedding the Economy in Society and Nature," *International Review of Sociology* 13, no. 2 (2003): 357–374; Gregory Baum, *Karl Polanyi on Ethics and Economics* (Toronto: McGill-Queen's University Press, 1996); Erik Ringmar, *The Mechanics of Modernity in Europe and East Asia: Institutional Origins of Social Change and Stagnation* (New York: Routledge, 2004); and J. Ron Stanfield, *The Economic Thought of Karl Polanyi: Lives and Livelihood* (London: Macmillan, 1986).

99 Mohai et al., "Environmental Justice"; Sze and London, "Environmental Justice at the Crossroads."

100 I use this term to denote grassroots social activism that privileges free market growth above social, community, and environmental well-being. Nancy Fraser began using the term in various talks and white papers in late 2012, but my use is quite distinct from her emancipatory connotation. I do not see markets as potential emancipators (even from gendered institutions or persistent poverty) but as part of a neoliberalized political-economic context in which individuals privilege free market systems despite the social and environmental degradation they have induced. While activism that uses markets and market-based logic may emancipate certain social groups (such as women) from the confines of the private sphere, it does not free them or anyone else from economic inequalities in neoliberalized, market-based societies.

101 Schlosberg, "Theorising Environmental Justice," 50.

102 Schlosberg, "Reconceiving Environmental Justice," 517.

103 Stephanie Malin, "There's No Real Choice but to Sign: Neoliberalization and Normalization of Hydraulic Fracturing on Pennsylvania Farmland," *Journal of Environmental Studies and Sciences* 4, no. 1 (2013): 17–27.

Chapter 2 Booms, Busts, and Bombs: Uranium's Economic and Environmental Justice History in the United States

1 Amundson, *Yellowcake Towns.*

2 Raye Ringholz, *Uranium Frenzy: Saga of the Nuclear West* (Logan: Utah State University Press, 2002).

3 Tom Zoellner, *Uranium: War, Energy, and the Rock That Shaped the World* (New York: Penguin, 2009).

4 Amundson, *Yellowcake Towns*; Ringholz, *Uranium Frenzy.*

5 Ringholz, *Uranium Frenzy.*

6 Amundson, *Yellowcake Towns.*

7 Joan Robinson, *Economics of Imperfect Competition* (London: Macmillan, 1933); Alan Manning, "The Real Thin Theory: Monopsony in Modern Labour Markets," *Labour Economics* 10, no. 2 (2003): 105–131.

8 Monopolies are economic arrangements in which many buyers have access to goods from only one seller, who may then set prices and control market dynamics.
9 Ringholz, *Uranium Frenzy*; Marie Templeton, *The Visionaries: First and Second Generation of the Piñon, Ute, and Nucla Areas* (Nucla, Colo.: Rimrocker Historical Society, 2010).
10 "The Mining Bug," *San Juan Record*, February 24, 1949, 4.
11 Ringholz, *Uranium Frenzy*, 71.
12 "Hope Uranium Hunt in West Equals Goldrush of Yesteryear," *San Juan Record*, December 6, 1951, 1.
13 "Extensive Uranium Program in Rich San Juan Basin Empire," *San Juan Record*, September 24, 1953,1.
14 "AEC Hikes Uranium Prices on Low-Grade Ores, Ups Bonuses," *San Juan Record*, March 8, 1951, 1; "Uranium Miners Paid Over One Million Dollars in Bonus," *San Juan Record*, March 6, 1953, 1.
15 For example, see "Energy in San Juan County," *San Juan Record*, February 17, 1955, insert.
16 Marie Templeton, *Naturita, Colorado: Where the Past Meets the Future* (Nucla, Colo.: Rimrocker Historical Society, 2002), 179.
17 Ibid., 134.
18 Amundson, *Yellowcake Towns*, 63.
19 Ringholz, *Uranium Frenzy*.
20 Ibid.
21 Ibid., 113.
22 Ibid., 126.
23 "Red Targets: America First?" *San Juan Record*, March 31, 1949, 2.
24 "Hoegh Says Every Family Needs a Fallout Shelter," *San Juan Record*, September 4, 1959, 2.
25 "Economic Highlights," *San Juan Record*, January 4, 1951, 2.
26 Ibid.
27 "What Other Editors Say: Uranium's Lure," *San Juan Record*, May 28, 1953, 4.
28 Ringholz, *Uranium Frenzy*.
29 Ibid., 214
30 "Senator Bennett Comes under Fire for 'Dragging Feet' in Uranium Mill Fight," *San Juan Record*, April 17, 1959, 1.
31 "Dictatorship in Democracy . . . ," *San Juan Record*, March 6, 1959, 1.
32 Ibid.
33 Ibid.
34 Amendments to the Atomic Energy Act invited private investment in the industry, while the Energy Reorganization Act of 1974 re-created federal regulatory bodies and institutions to make private investment increasingly seamless. The Energy Reorganization Act essentially divided the AEC into two separate entities: the Nuclear Regulatory Commission, charged with licensing activities pertaining to uranium and nuclear facilities; and the Department of Energy, charged with energy production and development activities outside of licensure, including the identification of potential environmental, health, safety, socioeconomic, institutional, and technology concerns related to nuclear technology development.
35 42 U.S.C., accessed October 8, 2014, http://www.law.cornell.edu/uscode/text/42/2011.

36 Amundson, *Yellowcake Towns*; Ringholz, *Uranium Frenzy*.
37 Sherry Cable and Charles Cable, *Environmental Problems, Grassroots Solutions: The Politics of Grassroots Environmental Politics* (New York: St. Martin's, 1994).
38 Templeton, *Naturita*, 179.
39 Ibid.
40 Ibid., 26.
41 Marie Templeton, *Standard Chemical Company: A Collection from the RHS* (Nucla, Colo.: Rimrocker Historical Society, 2007).
42 Templeton, *Naturita*, 179.
43 U.S. Bureau of the Census, *Profile of General Population and Housing Characteristics: 2010.*
44 City-Data.com, "Naturita, Colorado," accessed January 4, 2013, http://www.city-data.com/city/Naturita-Colorado.html.
45 U.S. Bureau of the Census, *Profile of General Population and Housing Characteristics: 2010*; City-Data.com, "Nucla, Colorado," accessed January 5, 2013, http://www.city-data.com/city/Nucla-Colorado.html.
46 Templeton, *Naturita*, 179.
47 I learned about these details during an in-depth interview and a Uravan site tour with members of the Rimrocker Historical Society. See also Amundson, *Yellowcake Towns*.
48 Union Carbide and Carbon Corporation, "Promise of a Golden Future" [advertisement], *Scientific American* (March 1953), http://blog.modernmechanix.com/promise-of-a-golden-future/..
49 Amundson, *Yellowcake Towns*; Ringholz, *Uranium Frenzy*.
50 ${}_{92}$*Ura* ${}_{23}$*Van*, accessed August 29, 2014, http://www.uravan.com.
51 U.S. Environmental Protection Agency, "Uravan Uranium Project (Union Carbide)," accessed June 9, 2014, http://www2.epa.gov/region8/uravan-uranium-project-union-carbide.
52 Ibid.
53 Hessler, "The Uranium Widows."
54 Jason Zasky, "Uravan, Colorado," *Failure Magazine*, November 20, 2012, http://failuremag.com/feature/article/uravan-colorado.
55 U.S. Agency for Toxic Substances and Disease Registry, *Preliminary Public Health Assessment*.
56 Ken Silver, "The Yellowed Archives of Yellowcake," *Public Health Reports* 111, no. 2 (1996): 116.
57 U.S. Agency for Toxic Substances and Disease Registry, *Preliminary Public Health Assessment*.
58 Lisa Church, "Mill Produces Cancer Rumors: Monticello Deals with Decades-Old Uranium Mill," *Salt Lake Tribune*, October 3, 2005, B1–B2.
59 As I will detail in chapter 3, the agency's report noted unusual health problems and high rates of breast cancer, lung cancer and other lung diseases, birth defects, and renal failure in Monticello. Yet both the agency and the EPA assured residents that no statistically significant link existed between their health outcomes and exposure to the mill site.
60 U.S. Agency for Toxic Substances and Disease Registry, *Preliminary Public Health Assessment*.
61 Amundson, *Yellowcake Towns*.

62 Ringholz, *Uranium Frenzy*, 80.
63 Ibid., 83.
64 Amundson, *Yellowcake Towns*, 86.
65 Ibid.
66 Ibid.
67 Ibid., 63.
68 Ibid.
69 Ibid. The mill added a new 3-million-dollar refinement practice and even signed a 300-million-dollar contract with the AEC to supply 27 million pounds of yellowcake between 1959 and 1961, making Moab's boom period longer than that of other uranium communities.
70 Ringholz, *Uranium Frenzy*.
71 The Uranium Reduction Company and later the Atlas Minerals Corporation created one of the largest uranium tailings waste sites in the United States: a 130-acre pile embedded in 400 acres of highly contaminated soil, situated about three miles north of Moab alongside the Colorado River. Because Atlas went bankrupt in 1998, the Department of Energy now acts as custodian and owner of the site and has overseen remediation of about one-third of the contamination under the legislated Uranium Mill Tailings Remediation Action Project. According to the local newspaper, remediation did not begin until 2007. In the meantime, the tailings pile sat uncovered and unsecured on the edge of the river for decades, which led to groundwater contamination, air pollution from tailings dust, and other types of contamination (Charli Engelhorn, "County Makes Another Plea for More Federal Funding for Uranium Mill Tailings Clean Up Project," *Moab Times-Independent*, August 30, 2012, http://www.moabtimes.com/view/full_story/19979353/article-County-makes-another-plea-for-more-federal-funding-for-uranium-mill-tailings-cleanup-project). Federal funding for completing the project has become increasingly precarious; already work on remediation has slowed as stimulus dollars have faded ("Moab UMTRA Project," *Grand County, Utah*, accessed January 25, 2013, http://www.moabtailings.org).
72 Amundson, *Yellowcake Towns*, 78.
73 For example, see Doug Brugge and Rob Goble, "The History of Uranium Mining and the Navajo People," *American Journal of Public Health* 92, no. 9 (2002): 1410–1419; Doug Brugge, Timothy Benally, and Esther Yazzie-Lewis, *The Navajo People and Uranium Mining* (Albuquerque: University of New Mexico Press, 2007); and Barbara Rose Johnston, Susan E. Dawson, and Gary E. Madsen, "Uranium Mining and Milling: Navajo Experiences in the American Southwest," in *The Energy Reader*, ed. Sherry Smith and Brian Frehner (Santa Fe, N.M.: SAR Press, 2007), 132–148.
74 Navajo Nation, "Facts at a Glance," accessed October 8, 2014, http://www.navajobusiness.com/fastFacts/Overview.htm.
75 Brugge et al., *The Navajo People*.
76 Johnston et al., "Uranium Mining and Milling."
77 Chris Shuey, *Uranium Exposure and Public Health in New Mexico and the Navajo Nation: A Literature Summary* (Albuquerque, N.M.: Southwest Research and Information Center, 2007).
78 Johnston et al., "Uranium Mining and Milling."
79 Ringholz, *Uranium Frenzy*, 19.

80 Ibid. This information is from Batie's later testimony in the *Begay v. United States* uranium exposure case.
81 Zoellner, *Uranium*.
82 Ringholz, *Uranium Frenzy*, 23.
83 Duncan Holaday, Wilfred David, and Henry Doyle, "An Interim Report of a Health Study of the Uranium Mills and Mines, 1952" (Salt Lake City, Utah: U.S. Public Health Service, Division of Occupational Health, May 1952).
84 Susan E. Dawson and Gary E. Madsen, "Worker Activism and Environmental Justice: The Black Lung and Radiation Compensation Programs," *Tulane Studies in Social Welfare* 21, no. 22 (2000): 209–228; Holaday et al., *An Interim Report*; Duncan Holaday, Victor Archer, and Frank Ludlin, "A Summary of United States Exposure Experiences in the Uranium Mining Industry," in *Symposium on Diagnosis and Treatment of Deposited Radionuclides*, ed. Henry Kornberg and William Norwood (Richmond, Wash.: Excerpta Medica Foundation, 1968), 451–456; Gary E. Madsen and Susan E. Dawson, "Unfinished Business: Radiation Exposure Compensation Act (RECA) for Post-1971 U.S. Uranium Underground Miners," *Journal of Health and Social Policy* 19, no. 4 (2005): 45–59; Gary E. Madsen, Susan E. Dawson, and Bryan R. Spykerman, "Perceived Occupational and Environmental Exposures: A Case Study of Former Uranium Millworkers," *Environment and Behavior* 28, no. 5 (1996): 571–590.
85 R. G. Beverly, "Topical Report WIN-114: Survey and Prevention Techniques for Control of Radioactivity Hazards at the Monticello Uranium Mill" (Washington, D.C.: National Lead Company, 1958).
86 Victor E. Archer, Joseph K. Wagoner, and Frank E. Lundin, Jr., "Cancer Mortality among Uranium Mill Workers," *Journal of Occupational and Environmental Medicine* 15, no. 1 (1973): 11–14; Richard J. Waxweiler, Victor E. Archer, Robert J. Roscoe, Arthur Watanabe, and Michael J. Thun, "Mortality Patterns among a Retrospective Cohort of Uranium Mill Workers," in *Epidemiology Applied to Health Physics* , ed. Gregg S. Wilkinson, (Albuquerque, N.M.: Proceedings of Health Physics Society, 1983), 428–435.
87 Victims of Mill Tailings Exposure survey, 2007. This is not a published document but was given to me by VMTE members.
88 "Petition Presented to VCA," *San Juan Record*, April 5, 1945, 8.
89 "Guest Editorial: And in the End We All Turn to It," *San Juan Record*, March 15, 1956, 2.
90 Ringholz, *Uranium Frenzy*.
91 Ibid.
92 Ibid.
93 See, for example, Carole Gallagher, *American Ground Zero: The Secret Nuclear War* (Cambridge, Mass.: MIT Press, 1993); and Ringholz, *Uranium Frenzy*.
94 Gallagher, *American Ground Zero*.
95 Brugge et al., *The Navajo People*; Silver, "The Yellowed Archives"; Richard W. Clapp, "Popular Epidemiology in Three Contaminated Communities," *Annals of the American Academy of Political and Social Science* 584, no. 1 (2002): 35–46; Malin and Petrzelka, "Left in the Dust."
96 Malin and Petrzelka, "Left in the Dust."
97 Brugge and Buchner, "Health Effects of Uranium: New Research Findings"; Johnston et al., "Uranium Mining and Milling."

98 Ringholz, *Uranium Frenzy*.

99 During his miller and miner studies in the 1950s, Holaday had developed a unit of measure he called a *working level*, which recorded the cumulative exposure of each worker to allow for easier calculations and comparisons across facilities. His recommended safe dosage was 1 working level per month (or 100 picocuries). The AEC did not follow that recommendation at the time, and debate still exists about correct threshold levels and working-level limits.

100 Ringholz, *Uranium Frenzy*.

101 Frank D. Gilliland, William C. Hunt, Marla Pardilla, and Charles R. Key, "Uranium Mining and Lung Cancer among Navajo Men in New Mexico and Arizona, 1969 to 1993," *Journal of Occupational and Environmental Medicine* 42, no. 3 (2000): 278–283; John Samet and David W Maple, "Diseases of Uranium Miners and Other Underground Miners Exposed to Radon," in *Environmental and Occupational Medicine*, ed. William Rom (Philadelphia: Lippincott-Raven, 1998), 1307–1315.

102 Brugge and Goble, "The History of Uranium Mining."

103 Shuey, *Uranium Exposure and Public Health*; Lynn E. Pinkerton, T. F. Bloom, M. J. Hein, and E. M. Ward, "Mortality among a Cohort of Uranium Mill Workers: An Update," *Occupational and Environmental Medicine* 61, no. 1 (2004): 57–64; Waxweiler et al., "Mortality Patterns"; Archer et al., "Cancer Mortality."

104 Shuey, *Uranium Exposure and Public Health*.

105 Stephanie Raymond-Whish et al., "Drinking Water with Uranium below the U.S. EPA Water Standard Causes Estrogen Receptor–Dependent Responses in Female Mice," *Environmental Health Perspectives* 115, no. 12 (2007): 1711–1716.

106 Brugge and Goble, "The History of Uranium."

107 Johnston et al., "Uranium Mining and Milling"; Dawson and Madsen, "Worker Activism"; Madsen et al., "Perceived Occupational and Environmental Exposures." A federally funded tribal health advocacy program called Navajo Nation's Community Health Representatives initially helped identify work-related health issues. Four other support groups also formed on the Nation (two each for miners and millers), and a fifth now provides assistance to families that are trying to make claims under the Radiation Exposure Compensation Act. A registry for uranium workers and their families has been established as the Navajo Office of Uranium Workers, and the Uranium Education Program at the Navajo Nation's Diné College helps translate technological and nuclear industry terms into accessible Navajo language and dialects so that people become more aware of the risks of exposure.

108 Ringholz, *Uranium Frenzy*.

109 Some of this remediation continues today—for instance, the enormous Atlas Mill tailings pile in Moab. Others, such as the 250-million-dollar remediation of two sites in Monticello, ended in the late 1990s and early 2000s. By 2006, Uravan had been entirely destroyed and the population relocated.

110 John D. Boice, Sarah S. Cohen, Michael T. Mumma, Bandana Chadda, and William J. Blot, "Mortality among Residents of Uravan, Colorado, Who Lived near a Uranium Mill, 1936–84," *Journal of Radiological Protection* 27, no. 3 (2007): 299.

Chapter 3 Lethal Legacies: Left in the Dust in Monticello, Utah

1 Portions of this chapter appeared in Malin and Petrzelka, "Left in the Dust"; and Stephanie A. Malin and Peggy Petrzelka, "Community Development among Toxic Tailings: An Interactional Case Study of Extralocal Institutions and Environmental Health," *Community Development* 43, no. 3 (2011): 379–392.

2 Ken Silver, "The Yellowed Archives of Yellowcake," *Public Health Report* 111 (1996): 116–127.

3 "This Man Needs Your Help," *San Juan Record*, January 16, 1952, 1.

4 "A-Bombs: Super-Duper," *San Juan Record*, February 3, 1949, 2.

5 Ibid.

6 "VCA Defense Plant about Completed," *San Juan Record*, September 10, 1942, 1.

7 "The Uranium Mill: The Heart of Our Economy," *San Juan Record*, August 18, 1955, 7.

8 Untitled article, *San Juan Record*, February 17, 1949, 10.

9 "Monticello Uranium Boom Creates Million Dollar Building Program," *San Juan Record*, June 23, 1955, 1; "Grand Openings Set Monticello's Pace for Expansion in U-ore Center," *San Juan Record*, April 28, 1953, 1.

10 "Petition Presented to VCA." In 1945, the Monticello Mill had briefly halted production when ownership shifted from the Vanadium Corporation of America to the AEC, which may explain why the fumes were more noticeable to residents after the mill resumed operations.

11 "Guest Editorial: And in the End We All Turn to It," *San Juan Record*, March 15, 1956, 2.

12 U.S. Agency for Toxic Substances and Disease Registry, *Preliminary Public Health Assessment.*

13 Ibid. More information is available at U.S. Environmental Protection Agency, "Utah Cleanup Sites," accessed October 10, 2014, http://www2.epa.gov/region8/monticello-mill-tailings-usdoe.

14 Victims of Mill Tailings Exposure, "The Monticello VMTE Committee," accessed June 8, 2014, https://sites.google.com/site/monticellovmte/home.

15 Brown, *Toxic Exposures*; Brown and Mikkelsen, *No Safe Place.*

16 Victims of Mill Tailings Exposure surveys, 1993, 2007.

17 "Monticello Cancer Study Points to Uranium Mill," *Association Press*, March 1, 2008, http://www.ksl.com/?nid=148&sid=2765888.

18 Victims of Mill Tailings Exposure surveys, 1993, 2007.

19 U.S. Agency for Toxic Substances and Disease Registry, *Preliminary Public Health Assessment.* The EPA finished its work in the late 1990s (though five property owners refused to allow their land to be remediated). In subsequent site checks, researchers have found contamination in the creek that runs through town and in the shallow alluvial aquifer under the creek. The town's water supply is upstream from the mill, but some people are concerned about future wells located down-river that might tap into the aquifer or worry that the aquifer will leak into a lower aquifer that is part of the town's water supply. Scientists have dissected local cattle and deer and monitored their organs for carcinogenic material they might have ingested near the site. While in 1997 the Agency for Toxic Substances and Disease

Registry found no more radionuclides in them than in control animals, officials were reluctant to rule out the possibility of eventual contamination. The agency has recommended continual monitoring of all unremediated properties, all agricultural products grown around the site, and all game hunted near the site as well as ongoing analysis of radon levels in homes.

20 Ibid.

21 Ibid.

22 Peer Review Committee of the Consortium for Risk Evaluation with Stakeholder Participation, "Review of the ATSDR Report Entitled 'Preliminary Public Health Assessment for Monticello Mill Tailings (DOE) Monticello, San Juan County, Utah, CERCLIS No. UT3890090035, December 28, 1995'" (Nashville, Tenn.: Consortium for Risk Evaluation with Stakeholder Participation, March 12, 1997).

23 Richard Clapp, "Popular Epidemiology in Three Contaminated Communities," *Annals of the American Academy of Political and Social Science* 584, no. 1 (2002): 35–46.

24 Ibid. Though Clapp mentions the meeting, he does not specify when it was held.

25 Lee Bennett, "Effects of the Cold War Still Felt in San Juan County: Mill Study Raises Questions and Answers," *San Juan Record*, February 12, 1997, 1, 13.

26 Ibid.

27 "Five AEC Houses Up for Sale," *San Juan Record*, May 9, 1963, 1.

28 "City Will Utilize Mill Tanks to Increase Water Storage Capacity," *San Juan Record*, August 8, 1963, 1.

29 "Final Notice: Public Must Respond by April 30, 1996, to Be Included in Property Cleanup," *San Juan Record*, April 15, 1996, 2.

30 Gallagher, *American Ground Zero*, 101.

31 Ibid., xxxii.

32 For example, see Holaday et al., "A Summary of United States Exposure Experiences."

33 U.S. Bureau of Land Management, staff report (Monticello, Utah, 1985), 1.

34 Ibid., 2.

35 Ibid.

36 "Five AEC Houses Up for Sale."

37 Joe Torres, letter to President George W. Bush, *San Juan Record*, March 7, 2001, 6.

38 Erin Brockovich was a legal clerk who was popularized in a movie depicting her instrumental role in fighting toxic contamination by Pacific Gas and Electric in California.

39 Cable and Cable, *Environmental Problems, Grassroots Solutions*.

40 Torres, letter.

41 "Monticello Cancer Study."

42 For missions statements and methodologies, see U.S. Agency for Toxic Substances and Disease Registry, "National Conversation on Public Health and Chemical Exposures, accessed September 8, 2014, http://www.atsdr.cdc.gov/nationalconversation.

43 Gillian Generoso, Stefanie Raymond-Whish, Karen Chase, and Cheryl A. Dyer, "Estrogenic Effects of Uranium in Human Breast Cancer Cells," paper delivered at the Northern Arizona University Research Symposium, Flagstaff, August 2007; Stefanie Raymond-Whish et al., "Drinking Water with Uranium below the U.S.

EPA Water Standard Causes Estrogen Receptor–Dependent Responses in Female Mice," *Environmental Health Perspectives* 115, no. 12 (2008): 1711–1716.

44 Susan M. Pinney et al., "Health Effects in Community Residents near a Uranium Plant at Fernald, Ohio, USA," *International Journal of Occupational Medicine and Environmental Health* 16, no. 2 (2003): 139–153.

Chapter 4 The Piñon Ridge Uranium Mill: A Transnational Corporation Comes Home

1 Energy Fuels, "Company Profile," accessed October 10, 2014, http://www.energyfuels.com/mobile/corporate Energy Fuels, "Energy Fuels, Inc. and Denison Mines Corp. Execute Definitive Arrangement Agreement," accessed September 11, 2014, http://www.energyfuels.com/news/index.php?&content_id=174.

2 Energy Fuels, "Company Profile."

3 Major managing institutions are the Bureau of Land Management, the Department of Energy, and the National Forest Service.

4 Colorado Department of Public Health and Environment, *Environmental Impact Analysis: Energy Fuels Piñon Ridge Uranium Mill Radioactive Materials License Approval* (Denver, 2011), http://recycle4colorado.ipower.com/EnergyFuels/application/envrpt/index.htm; "Montrose Country Socioeconomic Impact Study," EPS no. 19841, March 31, 2010, http://recycle4colorado.ipower.com/EnergyFuels/postap/10docs/100420socioecon.pdf; "Review of Piñon Ridge Environmental Report," March 11, 2010, http://recycle4colorado.ipower.com/EnergyFuels/postap/10docs/100420socioecon.pdf.

5 For example, see news coverage of the Piñon Ridge Mill's first rounds of the permitting process, where Glasier discussed the mill's technological and environmental advancements (Will Sands, "A Uranium Paradox," *Durango Telegraph*, November 18, 2010, http://www.durangotelegraph.com/index.cfm/archives/2010/november-18-2010/a-uranium-paradox).

6 See Energy Fuels' description of its commitment to environmental quality and operational transparency in the zoning and permitting processes at http://recycle4colorado.ipower.com/EnergyFuels/postap/10docs/100420socioecon.pdf. Also see "The Piñon Ridge Mill," accessed October 29, 2014, http://ulpeis.anl.gov/documents/dpeis/references/pdfs/Energy_Fuels_2012a.pdf.

7 Kamp, "Uranium and Health."

8 Information included descriptions of the mill's design; socioeconomic, environmental, and health reports; assays of surrounding geology, groundwater, and surface water; investigations into background radiation levels; and plans for decommissioning the mill (Colorado Department of Public Health and Environment, *Environmental Impact Analysis*).

9 States News Service, "Piñon Ridge Uranium Mill License Application Meets State Regulatory Requirements," May 1, 2013, at http://www.highbeam.com/doc/1G1-327593875.html.

10 Colorado Department of Public Health and Environment, *Environmental Impact Assessment*; "Montrose County Socioeconomic Impact Study."

11 I gathered this information from Energy Fuels' public meetings held before 2010. Also see Colorado Department of Public Health and Environment, *Environmental Impact Analysis.*

12 In dry form, yellowcake (U_3O_8) can be shipped to offsite conversion facilities. In its "natural" form, it is comprised of three different uranium isotopes: U-234 and U-238 (99.3 percent) and U-235 (0.7 percent). Once in the conversion facility, the ratios of these isotopes are changed to make the product more amenable to power production. At conversion facilities, the U_3O_8 is converted to UF_6 and then shipped to an enrichment facility. The goal of enrichment is to increase the proportion of U-235 up to 5 percent, which is better for use in nuclear reactors. Technical information in this section came primarily from Energy Fuels, "Company Profile"; and Colorado Department of Public Health and Environment, *Environmental Impact Analysis.*

13 Colorado Department of Public Health and Environment, *Environmental Impact Analysis*,. These claims are contested by opponent groups such as the Sheep Mountain Alliance.

14 "Risk Assessment for Proposed Uranium and Vanadium Mill at the Piñon Ridge Property," accessed October 11, 2014, https://www.colorado.gov/cdphedir/hm/Radiation/licenseapplication/rpt%281%29riskassessment.pdf; "Meteorology, Air Quality, and Climatology Report," August 29, 2009, http://recycle4colorado.ipower.com/EnergyFuels/preap/09docs/maqc/090827meteorology.pdf. Also see company responses to and comments about environmental and air-quality regulations at http://recycle4colorado.ipower.com/EnergyFuels/preap/09docs/index.htm.

15 "Environmental Impact Report: Summary of Impacts," accessed December 18, 2013, https://www.colorado.gov/cdphedir/hm/Radiation/licenseapplication/environmentalreport/sec5.pdf.

16 "Risk Assessment"; "Meteorology, Air Quality, and Climatology Report."

17 Energy Fuels asserts that this liner system will also act as a buffer for water aquifers after the mill's operations cease in about 2050. See "Risk Assessment."

18 I detail issues related to regulatory compliance in chapter 6.

19 Power, *A Socioeconomic Analysis.*

20 "Montrose Country Socioeconomic Impact Study"; Power, *A Socioeconomic Analysis*; "Socioeconomics Baseline for the New Piñon Ridge Uranium Mill" (Lakewood, Colo., June 8, 2009).

21 Power, *A Socioeconomic Analysis.*

22 This pattern emerged during my fieldwork. See also Hessler, "The Uranium Widows." These details show the complexity of people's relationships with corporations when these companies are part of an industry that people connect to and identify as part of their community social fabrics and personal histories. There is a complex interaction among a variety of social variables here, not singular corporate influence and economic identity, as is argued in works such as Bell and York, "Community Economic Identity."

23 After buying Union Carbide, Dow Chemical took ownership of the site.

24 Dallas Holmes, "Rimrockers Sign Uravan Lease," *San Miguel Basin Forum*, April 18, 2013, 1.

25 The coalition also argued that independent studies showed that the Piñon Ridge facility and its tailings ponds were not adequately separated from water aquifers, recalled uranium's harmful environmental health legacies and the boom-bust tendencies of the industry, and warned against reliance on technologies that require extensive risk (http://www.nirs.org/cdpheletter.pdf).

26 See "Denver District Court Throws Out License to Build Piñon Ridge Uranium Mill—Again," accessed October 11, 2014, http://www.sheepmountainalliance.org/category/uranium/pinon-ridge-mine.
27 Transcripts of public meetings at Nucla High School, sponsored by Colorado Department of Public Health and Environment, January and February 2010, http://recycle4colorado.ipower.com/EnergyFuels/postap/10docs/100217transcript.pdf. Other public meeting information is available at http://recycle4colorado.ipower.com/EnergyFuels/postap/10docs/index.htm.
28 Telluride Foundation, "Make More Possible," accessed September 11, 2014, http://www.telluridefoundation.org.
29 Telluride Foundation, "Paradox Community Development Initiative," accessed September 11, 2014, http://www.telluridefoundation.org/index.php?page=paradox-community-development-initiative-pcdi.
30 Telluride Foundation, "Press Releases," accessed September 11, 2014, https://www.telluridefoundation.org/index.php?mact=News,cntnt01,detail,0&cntnt01articleid=216&cntnt01returnid=27.
31 Telluride Foundation, "West End Community Resources."
32 Dorothy Kosich, "Energy Fuels, Titan Uranium Announce Friendly Takeover Bid," *Mineweb*, October 26, 2011, http://www.mineweb.com/mineweb/content/en/mineweb-uranium?oid=138247&sn=Detail.
33 Energy Fuels, "Company Profile." I gathered dates, timelines, and details during my lengthy interview with George Glasier in 2010.
34 David L. Naftz, Anthony J. Ranalli, Ryan C. Rowland, and Thomas M. Marston, *Assessment of Potential Migration of Radionuclides and Trace Elements from the White Mesa Uranium Mill to the Ute Mountain Ute Reservation and Surrounding Areas, Southeastern Utah* (Reston, Va.: U.S. Geological Survey, 2012), 146.
35 Transcripts of public meetings at Nucla High School; Energy Fuels, public information sessions (Nucla, Colo., 2010).
36 "Varca Ventures Announces Appointment of Mike Thompson As Interim Chief Operating Officer," *PR Newswire*, July 19, 2012, http://www.prnewswire.com/news-releases/varca-ventures-announces-appointment-of-mike-thompson-as-interim-chief-operating-officer-163008456.html.
37 Transcripts of public meetings at Nucla High School.

Chapter 5 "Just Hanging on by a Thread": Isolation, Poverty, and Social Dislocation

1 Portions of this chapter appear in Stephanie A. Malin, "When Is 'Yes to the Mill!' Environmental Justice: Interrogating Sites of Acceptance in Response to Energy Development," *Analyse and Kritik*, forthcoming.
2 Gary Nabhan, *The Colorado Plateau; An Orientation and Invocation* (Flagstaff: Northern Arizona University's Center for Sustainable Environments, 2002).
3 Ibid. Population density is based on measurement of the thirty-one counties with more than half their area in the ecoregion designated as the Colorado Plateau. Within this set of criteria, the population of the plateau is about 2,225,573 (U.S. Bureau of the Census, *Profile of General Population and Housing Characteristics: 2010*).

4 U.S. Bureau of the Census, *Profile of General Population and Housing Characteristics: 2010.*
5 Ibid.
6 The closest city with a million or more residents is Phoenix, Arizona, located about 380 miles away.
7 Bassett, "Distancing Rural Poverty." Also see Freudenburg and Gramling, "Linked to What?"; Smith et al., "Growth, Decline, Stability, and Disruption"; Freudenburg and Frickel. "Digging Deeper: Mining-Dependent Regions in Historical Perspective"; Lerner, *Diamond.*
8 Bassett, "Distancing Rural Poverty," 3–4.
9 See Rebecca J. W. Cline et al., "Community-Level Social Support Responses in a Slow-Motion Technological Disaster: The Case of Libby, Montana," *American Journal of Community Psychology* 46, nos. 1–2 (2010): 1–18.
10 Kuletz, *The Tainted Desert.*
11 Keeling, "'Born in an Atomic Test Tube."
12 For example, see Malin and Petrzelka, "Left in the Dust"; Malin and Petrzelka, "Community Development."
13 Critics argue that current demarcation of the poverty line does not adequately capture the extent of poverty in the United States. In their view, taking the amount of money it costs to buy a shopping cart's worth of food and multiplying it by three fails to capture important differences in poverty related to housing, transportation costs, and regional differences. For example, see the recent discussion of the controversy over poverty lines, poverty rates, and valid forms of measurement in the *Stanford Social Innovation Review* (Rourke L. O'Brien and David S. Pedulla, "Beyond the Poverty Line," *Stanford Social Innovation Review* 8, no. 4 [2010], http://www.ssireview.org/articles/entry/beyond_the_poverty_line).
14 U.S. Bureau of the Census, *Profile of General Population and Housing Characteristics: 2010.*
15 Ibid.; U.S. Bureau of the Census, "Naturita Town, CO."
16 U.S. Bureau of the Census, *Profile of General Population and Housing Characteristics: 2010*; U.S. Bureau of the Census, "Monticello City, Utah," *American Fact Finder*, accessed October 13, 2014, http://factfinder2.census.gov/faces/nav/jsf/pages/community_facts.xhtml#none.
17 For examples of crowd-sourced data, see City-Data.com, "Naturita, Colorado"; City-Data.com, "Nucla, Colorado."
18 U.S. Department of Agriculture, "Geography of Poverty," accessed June 9, 2014, http://www.ers.usda.gov/topics/rural-economy-population/rural-poverty-well-being/geography-of-poverty.aspx.
19 Although the subject is beyond the scope of this book and my data collection, I believe that the varying developmental trajectories of uranium communities deserve more extensive historical-comparative attention.
20 My question asked respondents to gauge their support for renewed uranium production and the new mill on a Likert scale from 1 to 5, with 5 "strongly in favor" and 1 "strongly against."
21 "Montrose County Socioeconomic Impact Study"; Power, *A Socioeconomic Analysis.*
22 Although I use pseudonyms for most of the people I interviewed, I've identified Don by his real name, with permission, because of his status as a public

figure. Not only is he well known regionally, but he has been profiled in Peter Hessler's "Dr. Don," *New Yorker*, September 26, 2011, http://www.newyorker.com/reporting/2011/09/26/110926fa_fact_hessler?currentPage=a.

23 Boice et al., "Mortality among Residents of Uravan, Colorado."

24 Transcripts of public meetings at Nucla High School.

25 I discuss the Cotter Mill extensively in chapter 7.

26 Joe Hanel, "Coloradoans Grapple with Promise, Threat of Uranium," *Durango Herald*, August 9, 2009, http://www.durangoherald.com/article/20090809/NEWS01/308099963/0/SEARCH/Coloradans-grapple-with-promise-threat-of-uranium.

27 Brugge is a Tufts University public health researcher who has done extensive research on Navajo's experiences with uranium and has worked as a consultant for the Sheep Mountain Alliance.

28 I reviewed these uranium studies in chapters 1 and 2. Environmental justice researchers have seen this pattern elsewhere—for example, in research on mercury poisoning among Native American populations (Kai T. Erikson, *A New Species of Trouble: The Human Experience of Modern Disasters* [New York: Norton, 1994]).

29 Cynthia Duncan, "Understanding Persistent Poverty: Social Class Context in Rural Communities," *Rural Sociology* 61, no. 1 (1996): 103–124.

Chapter 6 "Better Regs" in an Era of Deregulation: Neoliberalized Narratives of Regulatory Compliance

1 Richard J. Lazarus, "The Greening of America and the Graying of United States Environmental Law: Reflections on Environmental Law's First Three Decades in the United States," *Virginia. Environmental Law Journal* 20 (2001): 75–106.

2 For a rare comprehensive discussion of trends in U.S. environmental activism, see Daniel A. Mazmanian and Michael E. Kraft, *Toward Sustainable Communities: Transition and Transformations in Environmental Policy* (Cambridge, Mass.: MIT Press, 2009).

3 Lazarus, "The Greening of America."

4 U.S. Environmental Protection Agency, "EPA History," accessed June 7, 2014, http://www2.epa.gov/aboutepa/epa-history; Ann Fenton, *Toxic World: A Chronicle of Agency Discretion and Inaction* (Chicago: Loyola University Chicago, School of Law, 2008).

5 Lazarus, "The Greening of America"; Mazmanian and Kraft, *Toward Sustainable Communities;* Fenton, *Toxic World.* New regulations governed lead-based paint in homes, products used by children, and eventually gasoline; pesticide product labels; fuel economy standards; and pesticides such as DDT; they also banned federal loans to companies in violation of new legislation.

6 Cable and Cable, *Environmental Problems, Grassroots Solutions*; Fenton, *Toxic World.*

7 As a result of activist agitation, the Comprehensive Environmental Response, Compensation, and Liability Act (commonly known as the Superfund) was passed in 1980, empowering the EPA to prosecute the dirtiest companies.

8 Mazmanian and Kraft, *Toward Sustainable Communities.*

9 Ibid.; Harvey, *A Brief History of Neoliberalism;* Cable and Cable, *Environmental Problems, Grassroots Solutions*; Lazarus, "The Greening of America."

10 Mazmanian and Kraft, *Toward Sustainable Communities.*
11 Cable and Cable, *Environmental Problems, Grassroots Solutions.*
12 Lazarus, "The Greening of America."
13 For example, the National Environmental Protection Act impact assessments became mere cost-benefit analytical instruments that did not allow for inclusion of economic externalities such as potential death or disease from exposure to pollutants. The double movement did reappear, and gains were made in a few realms, including more rigid Clean Water Act restrictions passed in 1990, increased awareness of toxics and hazardous substances, and growing concern over global environmental problems. Still, neoliberal policy measures shaped an era of devolution.
14 For an extensive discussion of the power of neoliberal ideology in global environmental governance and definitions of sustainable development, see Michael Goldman, *Imperial Nature: The World Bank and Struggles for Social Justice in the Age of Globalization* (New Haven, Conn.: Yale University Press, 2005).
15 Fenton, *Toxic World.*
16 Mazmanian and Kraft, *Toward Sustainable Communities,* 15.
17 Thus far, strategies for action include increased private-public partnerships, global summits on climate change, and activism resisting environmental degradation in households, communities, and nations. Yet industrial activity continues to expand.
18 Cable and Cable, *Environmental Problems, Grassroots Solutions*; Allan Schnaiberg, David N. Pellow, and Adam Weinberg, "The Treadmill of Production and the Environmental State," *Environmental State under Pressure* 10 (2002): 15–32.
19 The role of corporations in shaping environmental policy and public perceptions of such policies has become contentious and consequential. Private entities such as Energy Fuels create allies through their donations to schools, hospitals, or even entire communities. At the same time spending cuts have truncated government's role, and "government regulation is increasingly curtailed around many environmental policies" (Schnaiberg et al., "The Treadmill of Production," 421).
20 Zoellner, *Uranium.*
21 Atomic Energy Act of 1946 (amended in 1954), P.L. 83-703, accessed October 11, 2014, http://pbadupws.nrc.gov/docs/ML1327/ML13274A489.pdf#page=23.
22 See U.S. Nuclear Regulatory Commission, "History," accessed September 17, 2014, http://www.nrc.gov/about-nrc/history.html.
23 Ringholz, *Uranium Frenzy.*
24 Amundson, *Yellowcake Towns;* Ringholz, *Uranium Frenzy.*
25 U.S. Nuclear Regulatory Commission, Office of General Counsel, "Nuclear Regulatory Legislation," NUREG-0980, 1, no. 10 (2013), http://pbadupws.nrc.gov/docs/ML1327/ML13274A489.pdf.
26 U.S. Nuclear Regulatory Commission, "Agreement State Program."
27 Given Colorado's status as an Agreement State, the project is not technically a federal one and will not require an environmental impact assessment under the National Environmental Protection Act.
28 Transcripts of public meetings at Nucla High School.
29 Sheep Mountain Alliance, "Hearings over Proposed Uranium Mill End, but Coloradans' Concerns over Potential Dangers, State Regulation Remain," November 15, 2012, http://www.sheepmountainalliance.org/uranium/hearings-over-proposed-uranium-mill-end-but-coloradans-concerns-over-potential-dangers-state-regulation-remain.

30 Kamp, "Uranium and Health."
31 U.S. Department of Labor, Mine Safety and Health Administration, "Mine Data Retrieval System," accessed September 17, 2014, http://www.msha.gov/drs/drshome.htm.
32 For example, see Uranium Watch, "Issues at White Mesa Mill (Utah)," accessed October 14, 2014, http://www.wise-uranium.org/umopwm.html.
33 According to critics, corporate self-monitoring and even third-party private monitoring for regulatory compliance can lead to lack of oversight, enforcement, and conflicts of interest. See Tim Bartley, "Certifying Forests and Factories: States, Social Movements, and the Rise of Private Regulation in the Apparel and Forest Products Fields." *Politics and Society* 31, no. 3 (2003): 433–464; Bartley, "Institutional Emergence in an Era of Globalization."
34 Colorado Department of the Treasury. "Constitutional Provisions," accessed June 7, 2014, http://www.colorado.gov/cs/Satellite/Treasury_v2/CBON/1251592160342.
35 Ibid. Spending limits are calculated using a complex formula that accounts for state population growth, federal rates of inflation, and other considerations.
36 Nicholas Johnson, Phil Oliff, and Erica Williams, "An Update on State Budget Cuts: At Least 46 States Have Imposed Cuts That Hurt Vulnerable Residents and the Economy" (Washington, D.C.: Center on Budget and Policy Priorities, February 9, 2011); Iris J. Lav and Erica Williams, "A Formula for Decline: Lessons from Colorado for States Considering TABOR" (Washington, D. C.: Center on Budget and Policy Priorities, March 15, 2010).
37 Via referendum C in 2005, taxpayers have since voted to amend some provisions and reduce revenue limits for a trial period between 2006 and 2010. However, the state is still feeling the effects of initial revenue reductions. Federal budget cuts have also exacerbated the shortcomings of regulatory enforcement in Colorado, even as the state budgets are evening out.
38 Placerville is about twenty miles away from Telluride. A site of gold, vanadium, uranium mining during the first and second uranium booms, the area is currently being remediated under the watch of the Colorado Department of Public Health and Environment.
39 When I ran crosstabs on those who believed that regulations were adequate with demographic variables in the survey instrument, I found no significant differences between respondents with faith in regulations and those who were skeptical of them. For example, in terms of household income, there was no significant difference between those with faith in regulations and those without ($\chi 2 = 9.784$, with a 0.460 significance level). My results show that concern about regulations does exist in pockets, even locally.
40 Popke, "The Spaces of Being In-Common," 243.
41 Noel Castree, "Neoliberalism and the Biophysical Environment: A Synthesis and Evaluation of the Research," *Environment and Society* 1 (2010): 11.

Chapter 7 Conclusions and Solutions: Social Sustainability and Localized Energy Justice

1 Cited in Fischer, "Clean Fuels Are a Social Panacea."
2 Thanks to Patricia Rose, professor of Africana studies at Brown University, for suggesting the term *insinuation* during a Cogut Center seminar.

3 For extended discussions of the transformative potential of social movement organizations, see Jaffee, *Brewing Justice*; Jaffee, "Weak Coffee"; and Nella Van Dyke, Sarah A. Soule, and Verta A. Taylor, "The Targets of Social Movements: Beyond a Focus on the State," *Research in Social Movements, Conflicts, and Change* 25 (2005): 27–51.
4 David N. Pellow, "Environmental Justice and the Political Process: Movements, Corporations, and the State," *Sociological Quarterly* 42, no. 1 (2001): 47–67.
5 Mann, *Playbook for Progressives*, x.
6 Jaffee, *Brewing Justice*, 112; Andrew Szasz, *Shopping Our Way to Safety: How We Changed from Protecting the Environment to Protecting Ourselves* (Minneapolis: University of Minnesota Press, 2010).
7 Trumpy, "Subject to Negotiation," 480.
8 Drew DeSilver, "U.S. Income Inequality, on Rise for Decades, Is Now Highest Since 1928," *FactTank*, accessed October 11, 2014, http://www.pewresearch.org/fact-tank/2013/12/05/u-s-income-inequality-on-rise-for-decades-is-now-highest-since-1928.
9 Malin, "There's No Real Choice but to Sign."
10 Sean Cockerham, "Colorado Cities' Fracking Bans Could Be Canary in a Coalmine," *McClatchyDC*, November 7, 2013, http://www.mcclatchydc.com/2013/11/07/207731/colorado-cities-fracking-bans.html. Interestingly, these moratoria have been attacked by Colorado Oil and Gas Association, the major lobbying group for the oil and gas industries in the state. For example, it successfully sued Fort Collins and forced the community to overturn its voter-initiated moratorium, though that ruling is currently under appeal.
11 Emily Guerin, "Navajo Nation's Purchase of a New Mexico Coalmine Is a Mixed Bag," *High County News*, January 7, 2014, http://www.hcn.org/blogs/goat/navajo-nations-purchase-of-a-new-mexico-coalmine-is-a-mixed-bag.
12 Chambers and Conway, *Sustainable Rural Livelihoods*, 6.
13 Brugge and Buchner, "Health Effects of Uranium"; Raymond-Whish et al., "Drinking Water with Uranium below the U.S. EPA Water Standard."
14 Social movement groups in sites of resistance suggest this as a potential solution as well.
15 Nucla-Naturita Chamber of Commerce, "Business Resources," accessed October 14, 2014, http://www.nucla-naturita.com/Business-Resources.html.
16 Clean Energy Collective, "San Miguel Power Installation," accessed October 14, 2014, http://easycleanenergy.com/communitysolarprojects.aspx?project=30ce76c2-432e-4576-bb10-be3f2d0d9ea8.
17 For example, see Kathryn J. Brasier et al., "Residents' Perception of Community and Environmental Impacts from Development of Natural Gas in the Marcellus Shale: A Comparison of Pennsylvania and New York Cases," *Journal of Rural Social Sciences* 26, no. 1 (2011): 32–61; Finewood and Stroup, "Fracking"; Shirley Stewart Burns, *Bringing down the Mountains The Impact of Mountaintop Removal on Southern West Virginia Communities* (Morgantown: West Virginia University Press, 2007); and Shannon Elizabeth Bell and Yvonne A. Braun, "Coal, Identity, and the Gendering of Environmental Justice Activism in Central Appalachia," *Gender and Society* 24, no. 6 (2010): 794–813.
18 For examples, see Mithra Moezzi and Loren Lutzenhiser, "What's Missing in Theories of the Residential Energy User" (Washington, D.C.: American Council

for an Energy-Efficient Economy, 2010): 7-207–7-221; Loren Lutzenhiser, "Social and Behavioral Aspects of Energy Use," *Annual Review of Energy and Environment* 18 (November 1993): 247–289; and Eugene A. Rosa, Gary A. Machlis, and Kenneth M. Keating, "Energy and Society," *Annual Review of Sociology* 14, no. 1 (1988): 149–172.

Appendix Research Methods and Data Collection

1 Earl R. Babbie, *The Basics of Social Research* (Belmont, Calif.: Wadsworth Thomson, 2005); Chambers and Conway, *Sustainable Rural Livelihoods.*
2 Malin and Petrzelka, "Left in the Dust."
3 Bruce L. Berg, *Qualitative Research Methods for the Social Sciences*, 8th ed. (Essex, U.K.: Pearson Education, 2014).
4 Don Dillman, Jolene D. Smyth, and Leah Melani Christian. *Internet, Mail, and Mixed Mode Surveys: The Tailored Design Method* (New York: Wiley, 2008).
5 Patricia Salant and Don Dillman. *How to Conduct Your Own Survey* (Hoboken, N.J.: Wiley, 1994).
6 I assumed an eighty-twenty split in community population characteristics, given Nucla's relative homogeneity, and strove for a 97 percent confidence interval.
7 I did not distribute a planned third wave of surveys because the Colorado Department of Public Health and Environment announced permit approval for construction of the Piñon Ridge Mill earlier than scheduled, which might have influenced householder responses. In addition, the *San Miguel Basin Forum* published an editorial about my survey, which included personal attacks on me. After a resident sent me this article, I responded with a letter to the editor addressing that misinformation. Nonetheless, I suspect the editorial may have had an impact on response rates.

Selected Bibliography

Achana, Francis T., and Joseph T. O'Leary. "The Transboundary Relationship between National Parks and Adjacent Communities." In *National Parks and Rural Development: Practice and Policy in the United States*, edited by Gary E. Machlis and Donald R. Field, 67–87. Washington, D.C.: Island Press, 2000.

Adaman, Fikret, Pat Devine, and Begum Ozkaynak. "Reinstituting the Economic Process: (Re)embedding the Economy in Society and Nature." *International Review of Sociology* 13, no. 2 (2003): 357–374.

Allen, Barbara L. *Uneasy Alchemy: Citizens and Experts in Louisiana's Chemical Corridor Disputes.* Cambridge, Mass.: MIT Press, 2003.

Amundson, Michael A. *Yellowcake Towns: Uranium Mining Communities in the American West.* Boulder: University Press of Colorado, 2002.

Archer, Victor E., Joseph K. Wagoner, and Frank E. Lundin, Jr. "Cancer Mortality among Uranium Mill Workers." *Journal of Occupational and Environmental Medicine* 15, no. 1 (1973): 11–14.

Babbie, Earl R. *The Basics of Social Research.* Belmont, Calif.: Wadsworth Thomson, 2005.

Bakker, Karen. *Privatizing Water: Governance Failure and the World's Urban Water Crisis.* Ithaca, N.Y.: Cornell University Press, 2010.

———. *An Uncooperative Commodity: Privatizing Water in England and Wales.* Oxford: Oxford University Press, 2003.

Bartley, Tim. "Certifying Forests and Factories: States, Social Movements, and the Rise of Private Regulation in the Apparel and Forest Products Fields." *Politics and Society* 31, no. 3 (2003): 433–464.

———. "Institutional Emergence in an Era of Globalization: The Rise of Transnational Private Regulation of Labor and Environmental Conditions." *American Journal of Sociology* 113, no. 2 (2007): 297–351.

Bassett, Debra Lyn. "Distancing Rural Poverty." *Georgetown Journal on Poverty Law and Policy* 13 (2006): 3–32.

Baum, Gregory. *Karl Polanyi on Ethics and Economics.* Toronto: McGill–Queen's University Press, 1996.

Bell, Shannon Elizabeth, and Yvonne A. Braun. "Coal, Identity, and the Gendering of Environmental Justice Activism in Central Appalachia." *Gender and Society* 24, no. 6 (2010): 794–813.

Bell, Shannon Elizabeth, and Richard York. "Community Economic Identity: The Coal Industry and Ideology Construction in West Virginia," *Rural Sociology* 75, no. 1 (2010): 111–143.

Berg, Bruce L. *Qualitative Research Methods for the Social Sciences.* 8th ed. Essex, U.K.: Pearson Education, 2014.

Beverly, R. G. "Topical Report WIN-114: Survey and Prevention Techniques for Control of Radioactivity Hazards at the Monticello Uranium Mill." Washington, D.C.: National Lead Company, 1958.

Block, Fred. "Polanyi's Double Movement and the Reconstruction of Critical Theory." *Revue Interventions Économiques* 38 (2008). http://interventionseconomiques.revues.org/274.

Boice, John D., Sarah S. Cohen, Michael T. Mumma, Bandana Chadda, and William J. Blot. "Mortality among Residents of Uravan, Colorado, Who Lived near a Uranium Mill, 1936–84." *Journal of Radiological Protection* 27, no. 3 (2007): 299.

Brasier, Kathryn J., Matthew R. Filteau, Diane K. McLaughlin, Jeffrey Jacquet, Richard C. Stedman, Timothy W. Kelsey, and Stephan J. Goetz. "Residents' Perceptions of Community and Environmental Impacts from Development of Natural Gas in the Marcellus Shale: A Comparison of Pennsylvania and New York Cases." *Journal of Rural Social Sciences* 26, no. 1 (2011): 32–61.

Brown, David Louis, Louis E. Swanson, and Alan W. Barton, eds. *Challenges for Rural America in the Twenty-First Century.* University Park: Pennsylvania State University Press, 2003.

Brown, Phil. *Toxic Exposures: Contested Illnesses and the Environmental Health Movement.* New York: Columbia University Press, 2007.

Brown, Phil, and Edwin Mikkelsen. *No Safe Place: Toxic Waste, Leukemia, and Community Action.* Berkeley: University of California Press, 1990.

Brown, Sandy, and Christy Getz. "Privatizing Farm Worker Justice: Regulating Labor through Voluntary Certification and Labeling." *Geoforum* 39, no. 3 (2008): 1184–1196.

Brugge, Doug, and Virginia Buchner. "Health Effects of Uranium: New Research Findings," *Review of Environmental Health* 26, no. 4 (2011): 231–249.

Brugge, Doug, Timothy Benally, and Esther Yazzie-Lewis. *The Navajo People and Uranium Mining.* Albuquerque: University of New Mexico Press, 2007.

Brugge, Doug, and Rob Goble. "The History of Uranium Mining and the Navajo People." *American Journal of Public Health* 92, no. 9 (2002): 1410–1419.

Bryant, Bunyan I., and Paul Mohai. *Race and the Incidence of Environmental Hazards: A Time for Discourse.* Boulder, Colo.: Westview, 1992.

Buğra, Ayşe, and Kaan Ağartan. *Reading Karl Polanyi for the Twenty-First Century: Market Economy As a Political Project.* New York: Palgrave Macmillan, 2007.

Bullard, Robert D. *Confronting Environmental Racism: Voices from the Grassroots.* Cambridge, Mass.: South End, 1993.

———. "A New 'Chicken-or-Egg' Debate: Which Came First—The Neighborhood, or the Toxic Dump?" *Workbook* 19, no. 2 (1994): 60–62.

———. "Solid Waste Sites and the Black Houston Community." *Sociological Inquiry* 53, no. 2–3 (1983): 273–288.

Burns, Shirley Stewart. *Bringing down the Mountains: The Impact of Mountaintop Removal on Southern West Virginia Communities.* Morgantown: West Virginia University Press, 2007.

Burtle, Adam. "Structural Violence." Accessed August 21, 2014. http://www.structuralviolence.org/structural-violence/.
Bury, Jeffrey. "Livelihoods in Transition: Transnational Gold Mining Operations and Local Change in Cajamarca, Peru." *Geographical Journal* 170, no. 1 (2004): 78–91.
Cable, Sherry, and Charles Cable. *Environmental Problems, Grassroots Solutions: The Politics of Grassroots Environmental Politics.* London: Worth, 1994.
Caniglia, Beth Schaefer, and JoAnn Carmin. "Scholarship on Social Movement Organizations: Classic Views and Emerging Trends." *Mobilization* 10, no. 2 (2005): 201–212.
Casanova, Pablo Gonzalez. "Internal Colonialism and National Development." *Studies in Comparative International Development* 1, no. 4 (1965): 27–37.
Castree, Noel. "Neoliberalising Nature: The Logics of Deregulation and Reregulation." *Environment and Planning* A40, no. 1 (2008): 131.
———. "Neoliberalising Nature: Processes, Effects, and Evaluations." *Environment and Planning* A40, no. 1 (2008): 153.
———. "Neoliberalism and the Biophysical Environment 2: Theorising the Neoliberalisation of Nature." *Geography Compass* 4, no. 12 (2010): 1734–1746.
Chambers, Robert, and Gordon Conway. *Sustainable Rural Livelihoods: Practical Concepts for the 21st Century.* Brighton, U.K.: Institute of Development Studies, 1991.
Clapp, Richard W. "Popular Epidemiology in Three Contaminated Communities." *Annals of the American Academy of Political and Social Science* 584, no. 1 (2002): 35–46.
Cline, Rebecca J. W., Heather Orom, Lisa Berry-Bobovski, Tanis Hernandez, C. Brad Black, Ann G. Schwartz, and John C. Ruckdeschel. "Community-Level Social Support Responses in a Slow-Motion Technological Disaster: The Case of Libby, Montana." *American Journal of Community Psychology* 46, nos. 1–2 (2010): 1–18.
Contested Illnesses Research Group. *Contested Illnesses: Citizens, Science, and Health Social Movements*, edited by Phil Brown, Rachel Morello-Frosch, and Stephen Zavestoski. Berkeley: University of California Press, 2012.
Cox, Robert W. "The Crisis in World Order and the Challenge to International Organization." *Cooperation and Conflict* 29, no. 2 (1994): 99–113.
Dale, Gareth. "Social Democracy, Embeddedness, and Decommodification: On the Conceptual Innovations and Intellectual Affiliations of Karl Polanyi." *New Political Economy* 15, no. 3 (2010): 369–393.
Dawson, Susan E., and Gary E. Madsen. "Worker Activism and Environmental Justice: The Black Lung and Radiation Compensation Programs." *Tulane Studies in Social Welfare* 21, no. 22 (2000): 209–228.
DeSilver, Drew. "U.S. Income Inequality, on Rise for Decades, Is Now Highest Since 1928." *FactTank.* Accessed October 11, 2014. http://www.pewresearch.org/fact-tank/2013/12/05/u-s-income-inequality-on-rise-for-decades-is-now-highest-since-1928.
Dillman, Don, Jolene D. Smyth, and Leah Melani Christian. *Internet, Mail, and Mixed Mode Surveys: The Tailored Design Method.* New York: Wiley, 2008.
Duncan, Cynthia M. "Understanding Persistent Poverty: Social Class Context in Rural Communities." *Rural Sociology* 61, no. 1 (1996): 103–124.
Erikson, Kai T. *A New Species of Trouble: The Human Experience of Modern Disasters.* New York: Norton, 1994.
Faber, Daniel, and Deborah McCarthy. "Neo-Liberalism, Globalization, and the Struggle for Ecological Democracy: Linking Sustainability and Environmental Justice." In *Just Sustainabilities: Development in an Unequal World*, edited by Julian Agyeman, Robert D. Bullard, and Bob Evans, 38–63. Cambridge, Mass: MIT Press, 2003.

Farmer, Paul E. *Pathologies of Power: Health, Human Rights, and the New War on the Poor.* 2nd ed. Berkeley: University of California Press, 2004.

Farmer, Paul E., Bruce Nizeye, Sara Stulac, and Salmaan Keshavjee. "Structural Violence and Clinical Medicine." *PLOS Medicine* 3, no. 10 (2006): e449.

Fenton, Ann. *Toxic World: A Chronicle of Agency Discretion and Inaction.* Chicago: Loyola University Chicago, School of Law, 2008.

Finewood, Michael H., and Laura J. Stroup. "Fracking and the Neoliberalization of the Hydro-Social Cycle in Pennsylvania's Marcellus Shale." *Journal of Contemporary Water Research and Education* 147, no. 1 (2012): 72–79.

Fletcher, Robert. "Neoliberal Environmentality: Towards a Poststructuralist Political Ecology of the Conservation Debate." *Conservation and Society* 8, no. 3 (2010): 171–181.

Foucault, Michel. *The Birth of Biopolitics: Lectures at the Collège de France, 1978–79*, edited by Michel Senellart, translated by Graham Burchell. Houndmills, U.K.: Palgrave Macmillan, 2008.

Freudenburg, William R. "Addictive Economies: Extractive Industries and Vulnerable Localities in a Changing World Economy." *Rural Sociology* 57, no. 3 (1992): 305–332.

Freudenburg, William R., and Scott Frickel. "Digging Deeper: Mining-Dependent Regions in Historical Perspective." *Rural Sociology* 59, no. 2 (1994): 266–288.

Freudenburg, William R., and Robert Gramling. "Linked to What? Economic Linkages in an Extractive Economy." *Society and Natural Resources* 11, no. 6 (1998): 569–586.

Frickel, Scott, and William R. Freudenburg. "Mining the Past: Historical Context and the Changing Implications of Natural Resource Extraction." *Social Problems* 43, no. 4 (1996): 444–466.

Gallagher, Carole. *American Ground Zero: The Secret Nuclear War.* Cambridge, Mass.: MIT Press, 1993.

Generoso, Gillian, Stefanie Raymond-Whish, Karen Chase, and Cheryl A. Dyer. "Estrogenic Effects of Uranium in Human Breast Cancer Cells." Paper delivered at the Northern Arizona University Research Symposium, Flagstaff, August 2007.

Gilliland, Frank D., William C. Hunt, Marla Pardilla, and Charles R. Key. "Uranium Mining and Lung Cancer among Navajo Men in New Mexico and Arizona, 1969 to 1993." *Journal of Occupational and Environmental Medicine* 42, no. 3 (2000): 278–283.

Goldman, Michael. *Imperial Nature: The World Bank and Struggles for Social Justice in the Age of Globalization.* New Haven, Conn.: Yale University Press, 2005.

Gramsci, Antonio. *Selections from the Prison Notebooks*, edited by Quintin Hoare and Geoffrey Noel Smith. New York: International, 1971.

Guthman, Julie. "Neoliberalism and the Making of Food Politics in California." *Geoforum* 39, no. 3 (2008): 1171–1183.

Hofrichter, Richard. *Toxic Struggles: The Theory and Practice of Environmental Justice.* Salt Lake City: University of Utah Press, 1993.

Harrison, Jill. "Abandoned Bodies and Spaces of Sacrifice: Pesticide Drift Activism and the Contestation of Neoliberal Environmental Politics in California." *Geoforum* 39, no. 3 (2008): 1197–1214.

———. *Pesticide Drift and the Pursuit of Environmental Justice.* Cambridge, Mass.: MIT Press, 2011.

Harvey, David. *A Brief History of Neoliberalism.* Oxford: Oxford University Press, 2005.

Hessler, Peter. "Dr. Don." *New Yorker*, September 26, 2011. http://www.newyorker.com/reporting/2011/09/26/110926fa_fact_hessler?currentPage=a.

———. "The Uranium Widows." *New Yorker*, September 13, 2010. http://www.newyorker.com/magazine/2010/09/13/the-uranium-widows.

Hettne, Björn. "Security and Peace in Post–Cold War Europe." *Journal of Peace Research* 28, no. 3 (1991): 279–294.

Heynen, Nik, and Harold A. Perkins. "Scalar Dialectics in Green: Urban Private Property and the Contradictions of the Neoliberalization of Nature." *Capitalism Nature Socialism* 16, no. 1 (2005): 99–113.

Heynen, Nik, and Paul Robbins. "The Neoliberalization of Nature: Governance, Privatization, Enclosure, and Valuation." *Capitalism Nature Socialism* 16, no. 1 (2005): 5–8.

Holaday, Duncan, Victor Archer, and Frank Ludlin. "A Summary of United States Exposure Experiences in the Uranium Mining Industry." In *Symposium on Diagnosis and Treatment of Deposited Radionuclides*, edited by Henry Kornberg and William Norwood, 451–456. Richmond, Wash.: Excerpta Medica Foundation,1968.

Holaday, Duncan, Wilfred David, and Henry Doyle. "An Interim Report of a Health Study of the Uranium Mills and Mines, 1952." Salt Lake City, Utah: U.S. Public Health Service, Division of Occupational Health, May 1952.

Holifield, Ryan. "Neoliberalism and Environmental Justice in the United States Environmental Protection Agency: Translating Policy into Managerial Practice in Hazardous Waste Remediation." *Geoforum* 35, no. 3 (2004): 285–297.

Hooks, Gregory, and Chad L. Smith. "The Treadmill of Destruction: National Sacrifice Areas and Native Americans." *American Sociological Review* 69, no. 4 (2004): 558–575.

———. "Treadmills of Production and Destruction Threats to the Environment Posed by Militarism." *Organization and Environment* 18, no. 1 (2005): 19–37.

Humphrey, Craig R., Gigi Berardi, Matthew S. Carroll, Sally Fairfax, Louise Fortmann, C. Geisler, T. G. Johnson, J. Kusel, R. G. Lee, and S. Macinko. "Theories in the Study of Natural Resource-Dependent Communities and Persistent Rural Poverty in the United States." *Persistent Poverty in Rural America*, special issue (1993): 136–172.

Hurley, Andrew. *Environmental Inequalities: Class, Race, and Industrial Pollution in Gary, Indiana, 1945–1980.* Chapel Hill: University of North Carolina Press, 1995.

Ishiyama, Noriko. "Environmental Justice and American Indian Tribal Sovereignty: Case Study of a Land-Use Conflict in Skull Valley, Utah." *Antipode* 35, no. 1 (2003): 119–139.

Jaffee, Daniel. *Brewing Justice: Fair Trade Coffee, Sustainability, and Survival.* Berkeley: University of California Press, 2007.

———. "Fair Trade Standards, Corporate Participation, and Social Movement Responses in the United States." *Journal of Business Ethics* 92, no. 2 (2010): 267–285.

———. "Weak Coffee: Certification and Co-optation in the Fair Trade Movement." *Social Problems* 59, no. 1 (2012): 94–116.

Johnson, Nicholas, Phil Oliff, and Erica Williams. "An Update on State Budget Cuts: At Least 46 States Have Imposed Cuts That Hurt Vulnerable Residents and the Economy." Washington, D.C.: Center on Budget and Policy Priorities, February 9, 2011.

Johnston, Barbara Rose, Susan E. Dawson, and Gary E. Madsen. "Uranium Mining and Milling: Navajo Experiences in the American Southwest." In *The Energy Reader*, edited by Sherry Smith and Brian Frehner, 132–148. Santa Fe, N.M.: SAR Press, 2007.

Keeling, Arn. "'Born in an Atomic Test Tube': Landscapes of Cyclonic Development at Uranium City, Saskatchewan." *Canadian Geographer* 54, no. 2 (2010): 228–252.

King, Brayden G., and Nicholas A. Pearce. "The Contentiousness of Markets: Politics, Social Movements, and Institutional Change in Markets." *Annual Review of Sociology* 36 (2010): 249–267.

Krannich, Richard S., and Albert E. Luloff. "Problems of Resource Dependency in U.S. Rural Communities." In *Progress in Rural Policy and Planning*, edited by Andrew Gilg, David Briggs, Robert Dilley, Owen Furuseth, and Geoff McDonald, 1:5–18. London: Bellhaven, 1991.

Krannich, Richard S., and Peggy Petrzelka. "Tourism and Natural Amenity Development: Real Opportunities." In *Challenges for Rural America in the Twenty-First Century*, edited by David Louis Brown, Louis E. Swanson, and Alan W. Barton, 190–199. University Park: Penn State University Press, 2003.

Kuletz, Valerie. "Appropriate/d Technology, Cultural Revival, and Environmental Activism." In *Appropriating Technology: Vernacular Science and Social Power*, edited by Ron Eglash, Jennifer L. Croissant, Giovanna diChiro, and Rayvon Fouche, 287 –305. Minneapolis: University of Minnesota Press. 2004.

———. "Invisible Spaces, Violent Places: Cold War Nuclear and Militarized Landscapes." In *Violent Environments*, edited by Nancy Lee Peluso and Michael Watts, 237–260. Ithaca, N.Y.: Cornell University Press, 2001.

———. *The Tainted Desert: Environmental Ruin in the American West.* New York: Routledge, 1998.

Lav, Iris J., and Erica Williams. "A Formula for Decline: Lessons from Colorado for States Considering TABOR." Washington, D.C.: Center on Budget and Policy Priorities, March 15, 2010.

Lazarus, Richard J. "The Greening of America and the Graying of United States Environmental Law: Reflections on Environmental Law's First Three Decades in the United States." *Virginia Environmental Law Journal* 20 (2001): 75–106.

Lerner, Steve. *Diamond: A Struggle for Justice in Louisiana's Chemical Corridor.* Cambridge, Mass.: MIT Press, 2005.

Lutzenhiser, Loren. "Social and Behavioral Aspects of Energy Use." *Annual Review of Energy and Environment* 18 (November 1993): 247–289.

Macfarlane, Allison, Michael Corradini, James K. Asselstine, and William G. Halsey. "Obstacles to Nuclear Power." *Bulletin of the Atomic Scientists* 63, no. 3 (2007): 24–25.

Madsen, Gary E., and Susan E. Dawson. "Unfinished Business: Radiation Exposure Compensation Act (RECA) for Post-1971 U.S. Uranium Underground Miners." *Journal of Health and Social Policy* 19, no. 4 (2005): 45–59.

Madsen, Gary E., Susan E. Dawson, and Bryan R. Spykerman. "Perceived Occupational and Environmental Exposures: A Case Study of Former Uranium Millworkers." *Environment and Behavior* 28, no. 5 (1996): 571–590.

Malin, Stephanie. "There's No Real Choice but to Sign: Neoliberalization and Normalization of Hydraulic Fracturing on Pennsylvania Farmland." *Journal of Environmental Studies and Sciences* 4, no. 1 (2013): 17–27.

Malin, Stephanie A., and Peggy Petrzelka. "Community Development among Toxic Tailings: An Interactional Case Study of Extralocal Institutions and Environmental Health." *Community Development* 43, no. 3 (2011): 379–392.

———. "Left in the Dust: Uranium's Legacy and Victims of Mill Tailings Exposure in Monticello, Utah." *Society and Natural Resources* 23, no. 12 (2010): 1187–1200.

Mann, Eric. *Playbook for Progressives: 16 Qualities of the Successful Organizer.* Boston: Beacon, 2011.

Manning, Alan. "The Real Thin Theory: Monopsony in Modern Labour Markets." *Labour Economics* 10, no. 2 (2003): 105–131.

Mansfield, Becky. "Privatization: Property and the Remaking of Nature–Society Relations." *Antipode* 39, no. 3 (2007): 393–405.

Mazmanian, Daniel A., and Michael E. Kraft. *Toward Sustainable Communities: Transition and Transformations in Environmental Policy.* Cambridge, Mass.: MIT Press, 2009.

McCarthy, James. "Commons As Counterhegemonic Projects." *Capitalism Nature Socialism* 16, no. 1 (2005): 9–24.

———. "Devolution in the Woods: Community Forestry As Hybrid Neoliberalism." *Environment and Planning* A37, no. 6 (2005): 995–1014.

Moezzi, Mithra, and Loren Lutzenhiser. "What's Missing in Theories of the Residential Energy User." Washington, D.C.: American Council for an Energy-Efficient Economy: Washington, D.C., 2010.

Mohai, Paul, David Pellow, and J. Timmons Roberts. "Environmental Justice." *Annual Review of Environment and Resources* 34 (2009): 405–430.

Murray, Warwick E. "The Neoliberal Inheritance: Agrarian Policy and Rural Differentiation in Democratic Chile." *Bulletin of Latin American Research* 21, no. 3 (2002): 425–441.

Nabhan, Gary. *The Colorado Plateau; An Orientation and Invocation.* Flagstaff: Northern Arizona University, Center for Sustainable Environments, 2002.

Naftz, David L., Anthony J. Ranalli, Ryan C. Rowland, and Thomas M. Marston. *Assessment of Potential Migration of Radionuclides and Trace Elements from the White Mesa Uranium Mill to the Ute Mountain Ute Reservation and Surrounding Areas, Southeastern Utah*, Reston, Va.: U.S. Geological Survey, 2012.

O'Brien, Rourke L., and David S. Pedulla. "Beyond the Poverty Line." *Stanford Social Innovation Review* 8, no. 4 (2010). http://www.ssireview.org/articles/entry/beyond_the_poverty_line.

Peck, Jamie, and Adam Tickell. "Neoliberalizing Space." *Antipode* 34, no. 3 (2002): 380–404.

Peer Review Committee of the Consortium for Risk Evaluation with Stakeholder Participation. "Review of the ATSDR Report Entitled 'Preliminary Public Health Assessment for Monticello Mill Tailings (DOE) Monticello, San Juan County, Utah, CERCLIS No. UT3890090035, December 28, 1995.'" Nashville, Tenn.: Consortium for Risk Evaluation with Stakeholder Participation, March 12, 1997.

Pellow, David N. "Environmental Justice and the Political Process: Movements, Corporations, and the State." *Sociological Quarterly* 42, no. 1 (2001): 47–67.

———. *Garbage Wars: The Struggle for Environmental Justice in Chicago.* Cambridge, Mass.: MIT Press, 2002.

Peluso, Nancy Lee, Craig R. Humphrey, and Louise P. Fortmann. "The Rock, the Beach, and the Tidal Pool: People and Poverty in Natural Resource–Dependent Areas." *Society and Natural Resources* 7, no. 1 (1994): 23–38.

Perreault, Thomas. "From the Guerra del Agua to the Guerra del Gas: Resource Governance, Neoliberalism, and Popular Protest in Bolivia." *Antipode* 38, no. 1 (2006): 150–172.

Pinkerton, Lynn E., T. F. Bloom, M. J. Hein, and E. M. Ward. "Mortality among a Cohort of Uranium Mill Workers: An Update." *Occupational and Environmental Medicine* 61, no. 1 (2004): 57–64.

Pinney, Susan M., Ronald W. Freyberg, Gail H. Levine, Donald E. Brannen, Lynn S. Mark, James M. Nasuta, Colleen D. Tebbe, Jeanette M. Buckholz, and Robert Wones. "Health Effects in Community Residents near a Uranium Plant at Fernald, Ohio, USA." *International Journal of Occupational Medicine and Environmental Health* 16, no. 2 (2003): 139–153.

Polanyi, Karl. *The Great Transformation: The Political and Economic Origins of Our Time.* Boston: Beacon, 1944.

Popke, Jeff. "The Spaces of Being In-Common: Ethics and Social Geography." In *The Sage Handbook of Social Geographies*, edited by Susan Smith, Rachel Pain, Sallie Marston, and John Paul Jones, 435–454. London: Sage, 2010.

Portes, Alejandro, and Kelly Hoffman. "Latin American Class Structures: Their Composition and Change during the Neoliberal Era." *Latin American Research Review* 38, no. 1 (2003): 41–82.

Power, Tom. *A Socioeconomic Analysis of the Impact of the Proposed Piñon Ridge Uranium Mill Project on Western Mesa, Montrose, and San Miguel Counties, Colorado.* Missoula, Mont.: Power Consulting, January 2011.

Prudham, Scott. "Poisoning the Well: Neoliberalism and the Contamination of Municipal Water in Walkerton, Ontario." *Geoforum* 35, no. 3 (2004): 343–359.

Pulido, Laura. "A Critical Review of the Methodology of Environmental Racial Research." *Antipode* 28, no. 2 (1996): 142–159.

Randles, Sally. "Issues for a Neo-Polanyian Research Agenda in Economic Sociology." In *Karl Polanyi: New Perspectives on the Place of the Economy in Society*, edited by Mark Harvey, Ronnie Ramlogan, and Sally Randles. 409–434. Manchester, U.K.: Manchester University Press, 2007.

Raymond-Whish, Stefanie, Loretta P. Mayer, Tamara O'Neal, Alisyn Martinez, Marilee A. Sellers, Patricia J. Christian, Samuel L. Marion, Carlyle Begay, Catherine R. Propper, and Patricia B. Hoyer. "Drinking Water with Uranium below the U.S. EPA Water Standard Causes Estrogen Receptor–Dependent Responses in Female Mice." *Environmental Health Perspectives* 115, no. 12 (2007): 1711–1716.

Ringholz, Raye. *Uranium Frenzy: Saga of the Nuclear West.* Logan: Utah State University Press, 2002.

Ringmar, Erik. *The Mechanics of Modernity in Europe and East Asia: Institutional Origins of Social Change and Stagnation.* New York: Routledge, 2004.

Roberts, J. Timmons, and Melissa M. Toffolon-Weiss. *Chronicles from the Environmental Justice Frontline.* Cambridge, U.K.: Cambridge University Press, 2001.

Robinson, Joan. *Economics of Imperfect Competition.* London: Macmillan, 1933.

Rosa, Eugene A., Gary E. Machlis, and Kenneth M. Keating. "Energy and Society." *Annual Review of Sociology* 14, no. 1 (1988): 149–172.

Salant, Patricia, and Don Dillman. *How to Conduct Your Own Survey.* Hoboken, N.J.: Wiley, 1994.

Samet, John, and David W Maple. "Diseases of Uranium Miners and Other Underground Miners Exposed to Radon," in *Environmental and Occupational Medicine*, edited by William Rom, 1307–1315. Philadelphia: Lippincott-Raven, 1998.

Schlosberg, David. "Reconceiving Environmental Justice: Global Movements and Political Theories." *Environmental Politics* 13, no. 3 (2004): 517–540.

———. "Theorising Environmental Justice: The Expanding Sphere of a Discourse." *Environmental Politics* 22, no. 1 (2013): 37–55.

Schnaiberg, Allan, David N. Pellow, and Adam Weinberg. "The Treadmill of Production and the Environmental State." *Environmental State under Pressure* 10 (2002): 15–32.

Shuey, Chris. *Uranium Exposure and Public Health in New Mexico and the Navajo Nation: A Literature Summary.* Albuquerque, N.M.: Southwest Research and Information Center, 2007.

Silver, Beverly J., and Giovanni Arrighi. "Polanyi's 'Double Movement': The Belle Époques of British and U.S. Hegemony Compared." *Politics and Society* 31, no. 2 (2003): 325–355.

Silver, Ken. "The Yellowed Archives of Yellowcake." *Public Health Reports* 111, no. 2 (1996): 116.

Smith, Michael D., Richard S. Krannich, and Lori M. Hunter. "Growth, Decline, Stability, and Disruption: A Longitudinal Analysis of Social Well-Being in Four Western Rural Communities." *Rural Sociology* 66, no. 3 (2001): 425–450.

Smith, Neil. *Uneven Development: Nature, Capital, and the Production of Space.* Athens: University of Georgia Press, 2008.

Stanfield, J. Ron. *The Economic Thought of Karl Polanyi: Lives and Livelihood.* London: Macmillan, 1986.

Stedman, Richard C. "Places of Escape: Second-Home Meanings in Northern Wisconsin, USA." In *Multiple Dwelling and Tourism: Negotiating Place, Home, and Identity*, edited by Norman McIntyre, Daniel Williams, and Kevin McHugh, 129–144. Oxfordshire, U.K.: CABI Press, 2006.

Szasz, Andrew, and Michael Meuser. "Environmental Inequalities: Literature Review and Proposals for New Directions in Research and Theory." *Current Sociology* 45, no. 3 (1997): 99–120.

Szasz, Andrew. *Shopping Our Way to Safety: How We Changed from Protecting the Environment to Protecting Ourselves.* Minneapolis: University of Minnesota Press, 2010.

Sze, Julie, and Jonathan K. London. "Environmental Justice at the Crossroads." *Sociology Compass* 2, no. 4 (2008): 1331–1354.

Taylor, Dorceta E. "The Rise of the Environmental Justice Paradigm: Injustice Framing and the Social Construction of Environmental Discourses." *American Behavioral Scientist* 43, no. 4 (2000): 508–580.

Templeton, Marie. *Naturita, Colorado: Where the Past Meets the Future.* Nucla, Colo.: Rimrocker Historical Society, 2002.

———. *Standard Chemical Company: A Collection from the RHS.* Nucla, Colo.: Rimrocker Historical Society, 2007.

———. *The Visionaries: First and Second Generation of the Piñon, Ute, and Nucla Areas.* Nucla, Colo.: Rimrocker Historical Society, 2010.

Trumpy, Alexa J. "Subject to Negotiation: The Mechanisms behind Co-Optation and Corporate Reform." *Social Problems* 55, no. 4 (2008): 480–500.

United Church of Christ Commission for Racial Justice. *Toxic Wastes and Race in the United States: A National Report on the Racial and Socio-Economic Characteristics of Communities with Hazardous Waste Sites.* New York: Public Data Access, 1987.

U.S. Agency for Toxic Substances and Disease Registry. "National Conversation on Public Health and Chemical Exposures." Accessed September 8, 2014. http://www.atsdr.cdc.gov/nationalconversation.

———. *Preliminary Public Health Assessment for Monticello Mill Tailings (DOE) and Monticello Radioactively Contaminated Properties (aka Monticello Vicinity Properties), Monticello, San Juan County, Utah.* Atlanta: Department of Health and Human Services, 1997.

U.S. General Accounting Office. *Siting of Hazardous Waste Landfills and Their Correlation with Racial and Economic Status of Surrounding Communities.* Gaithersburg, Md.: U.S. General Accounting Office, 1983.

Valdivia, Gabriela. "On Indigeneity, Change, and Representation in the Northeastern Ecuadorian Amazon." *Environment and Planning* A37, no. 2 (2005): 285–303.

Van Dyke, Nella, Sarah A. Soule, and Verta A. Taylor. "The Targets of Social Movements: Beyond a Focus on the State." *Research in Social Movements, Conflicts, and Change* 25 (2005): 27–51.

Walls, David S. "Internal Colony or Internal Periphery? A Critique of Current Models and an Alternative Formulation." In *Colonialism in Modern America: The Appalachian Case*, edited by Helen M. Lewis, Linda Johnson, and Don Askins, 319–349. Boone, N.C.: Appalachian Consortium Press, 1978.

Waxweiler, Richard J., Victor E. Archer, Robert J. Roscoe, Arthur Watanabe, and Michael J. Thun. "Mortality Patterns among a Retrospective Cohort of Uranium Mill Workers." In *Epidemiology Applied to Health Physics*, edited by Gregg S. Wilkinson, 428–435. Albuquerque, N.M.: Proceedings of Health Physics Society, 1983.

Zoellner, Tom. *Uranium: War, Energy, and the Rock That Shaped the World.* New York: Penguin, 2009.

Index

Page numbers followed by an f indicate a figure; those by a t, a table.

About the Author

STEPHANIE A. MALIN is an environmental sociologist specializing in rural community, governance, and political economies of natural resource extraction and energy development. Her main interests include environmental justice, environmental health, social mobilization, and the social-environmental effects of market-based economies. Malin earned a Ph.D. in sociology at Utah State University and completed a Mellon Foundation postdoctoral fellowship at Brown University. Currently she is an assistant professor in the Department of Sociology at Colorado State University. From her lovely base in Fort Collins, she continues her research on uranium production's history and social effects. She is also actively studying unconventional oil and gas development and its relationship to quality of life, notions of environmental justice, water rights, health and stress, and social activism. In her precious free time, she enjoys biking, hiking, camping, and traveling with her husband, family, friends, and trusty rescue dog.